T0135648

ABSORPTION SPECTROSCOPY STUDIES IN LOW PRESSURE NON EQUILIBRIUM MOLECULAR PLASMAS USING TUNABLE INFRARED DIODE LASERS

INAUGURAL DISSERTATION

zur

Erlangung des akademischen Grades

doctor rerum naturalium (Dr. rer. nat.)

an der Mathematisch-Naturwissenschaftliche Fakultät

der

Ernst-Moritz-Arndt-Universität Greifswald

vorgelegt von

Frank Hempel

geboren am 01. November 1971

in Greifswald

Greifswald, 03. Februar 2003

Bibliografische Information Der Deutschen Bibliothek

Die Deutsche Bibliothek verzeichnet diese Publikation in der Deutschen
Nationalbibliografie; detaillierte bibliografische Daten sind im Internet über
http://dnb.ddb.de abrufbar.

ISBN 3-8325-0262-9

Logos Verlag Berlin
Comeniushof, Gubener Str. 47,
10243 Berlin
Tel.: +49 030 42 85 10 90
Fax: +49 030 42 85 10 92
INTERNET: http://www.logos-verlag.de

Dekan: Prof. Dr. J.-P. Hildebrandt

1. Gutachter: Prof. Dr. J. P. F. Conrads

2. Gutachter: Dr. P. B. Davies

Tag der Promotion: 13. Mai 2003

Content

1   Introduction 5

2   Diagnostics of Molecular Plasmas...9
    2.1   Invasive Methods 9
    2.2   Non-Invasive Methods 11

3   Tunable Diode Laser Absorption Spectroscopy 19
    3.1   Diode Lasers 20
    3.2   Concentration Measurements by Direct Absorption Spectroscopy.......................21
    3.3   Derivative Spectroscopy...23
    3.4   Line Identification and Spectral Calibration.................................24

4   Plasma Chemistry 27

5   Investigations in Planar Microwave Discharges 31
    5.1   Introduction..31
    5.2   Power Transfer in Microwave Discharges.....................................32
    5.3   The Planar Microwave Discharge 36
        5.3.1   Fundamentals..36
        5.3.2   Properties..39
    5.4   Experimental System..41
        5.4.1   Discharge Reactor System..41
        5.4.2   Tunable Diode Laser Arrangement...45
        5.4.3   Rapid Scan Software...46
    5.5   Investigations of Hydrogen-Nitrogen Plasmas Containing Methane or Methanol 48
        5.5.1   Introduction...48
        5.5.2   Experimental Results...50
            5.5.2.1   General features..50
            5.5.2.2   Degree of Hydrocarbon Precursor Dissociation................54
            5.5.2.3   Mass Balance Considerations...55
            5.5.2.4   Main Products of the Discharge (HCN and NH₃)................57
            5.5.2.5   The Methyl radical (CH₃)..60
            5.5.2.6   Other Carbon containing products......................................63
            5.5.2.7   Fragmentation and Conversion rates..................................65
        5.5.3   Plasma Chemistry Modelling for Hydrogen-Nitrogen-Methane Discharges 68
            5.5.3.1   The Chemical Reaction Kinetics Model................................68
            5.5.3.2   Results..76
    5.6   Spectroscopic Investigations of the CN Free Radical....................79
        5.6.1   Experimental 80
        5.6.2   Results..81
        5.6.3   Discussion 85

6   Investigations in DC discharges **89**

6.1   Investigations of the Carbon Dioxide Conversion Chemistry in a Low Pressure
Glow Discharge 89
  6.1.1   Introduction *89*
  6.1.2   Experimental *90*
  6.1.3   Results and Discussion *93*
    6.1.3.1   Investigation of the Temporal Development of CO and CO$_2$ *93*
    6.1.3.2   Modelling *98*

6.2   HCl Concentration Measurements in Pulsed H$_2$-Ar-N$_2$-TiCl$_4$ DC Plasmas 101
  6.2.1   Introduction *101*
  6.2.2   Experimental *101*
  6.2.3   Results and Discussion *102*

7   Accuracy and Limitations **105**

8   Summary and Conclusions **107**

9   References **113**

10   Appendix **127**

  10.1   List of additional figures 127

Abstract

Tunable infrared diode laser absorption spectroscopy (TDLAS) has been applied to investigate the chemical kinetics in reactive discharges. It was used to detect the methyl radical and nine stable molecules, CH_4, CH_3OH, C_2H_2, C_2H_4, C_2H_6, NH_3, HCN, CH_2O and C_2N_2, in H_2-Ar-N_2 microwave plasmas containing up to 7.2 % of methane or methanol, under both flowing and static conditions. The degree of dissociation of the hydrocarbon precursor molecules varied between 20 and 97 %. The methyl radical concentration was found to be in the range 10^{12} to 10^{13} molecules cm^{-3}. By analysing the temporal development of molecular concentrations under static conditions it was found that HCN and NH_3 are the final products of plasma chemical conversion. The fragmentation rates of methane and methanol ($R_F(CH_4)$= 2–7 x 10^{15} molecules J^{-1}, $R_F(CH_3OH)$= 6–9 x 10^{15} molecules J^{-1}) and the respective conversion rates to methane, hydrogen cyanide and ammonia ($R_{Cmax}(CH_4)$= 1.2 x 10^{15} molecules J^{-1}, $R_{Cmax}(HCN)$= 1.3 x 10^{15} molecules J^{-1}, $R_{Cmax}(NH_3)$= 1 x 10^{14} molecules J^{-1}) have been determined for different hydrogen to nitrogen concentration ratios. An extensive model of the chemical reactions involved in the H_2-N_2-Ar-CH_4 plasma has been developed. Model calculations were performed by including 22 species, 145 chemical reactions and appropriate electron impact dissociation rate coefficients. The results of the model calculations showed satisfactory agreement between calculated and measured concentrations. The most likely main chemical pathways involved in these plasmas are discussed and an appropriate reaction scheme is proposed. Based on the model calculations the concentrations of non-measured species like CH_2 or NH_2 have been predicted.

In addition, spectroscopic investigations of P- and R-branch lines of the fundamental bands of $^{12}C^{14}N$ and $^{13}C^{14}N$ in their ground ($^2\Sigma^+$) electronic state have been performed at high resolution by tunable diode laser absorption spectroscopy. The radicals were generated in microwave plasmas containing methane with varying proportions of N_2 and H_2. From a fit to the spectra the origins of the fundamental bands of the two isotopomers were determined to be 2042.42104(84) cm^{-1} and 2000.08470(30) cm^{-1}. The main product detected in the plasma was HCN. It showed concentrations which are about three orders of magnitude higher than that of CN.

Moreover, the time and spatial dependence of the chemical conversion of CO_2 to CO were studied in a closed glow discharge reactor (p = 50 Pa, I = 2...30 mA) consisting of a small plasma zone and an extended stationary afterglow. Tunable infrared diode laser absorption spectroscopy has been applied to determine the absolute ground state concentrations of CO and CO_2. After a certain discharge time the concentrations of both species were observed to come into equilibrium. The spatial dependence of the equilibrium CO concentration in the afterglow was found to vary by less than 10 %. The feed gas was converted to CO more predominantly between 45 % and 60 % with increasing discharge current. The formation time of the stable gas composition decreased with increasing current too. For currents higher than 10 mA the conversion rate of CO_2 to CO was estimated to be 1×10^{13} molecules J^{-1}. Based on the experimental results a model of the CO_2 conversion chemistry has also been established for this type of discharge. The calculated and measured temporal developments of species concentrations showed a satisfactory agreement for various discharge currents.

Lastly, infrared tunable diode laser absorption spectroscopy has been used to analyse the fragmentation of $TiCl_4$ into HCl in pulsed H_2-Ar-N_2 dc plasmas (p= 2 mbar). At small $TiCl_4$ admixtures (0.04-0.31 %) HCl concentrations of 2-5 $\times 10^{14}$ molecules cm^{-3} were measured (current density: 0.6-1.15 mA cm^2). A nearly complete conversion of Cl into HCl was found at $TiCl_4$ admixtures below 0.2 %.

1   Introduction

The volume chemistry of low-pressure, non-equilibrium, molecular plasmas is due to their favourable properties, of growing interest not only in the field of plasma processing but also in basic research. This type of plasmas, which can be excited without using electrodes to couple microwave energy into the reaction chamber, is characterized by a high degree of dissociation of the precursor gases and high chemical reactivity due to the large concentrations of transient or stable chemically active neutral species present. The investigation of the plasma chemistry and kinetics requires detailed knowledge of the main plasma parameters, which can be obtained by appropriate diagnostic techniques. The online monitoring of transient or stable plasma reaction products in chemical reactors, in particular the measurement of their ground state concentrations, is the key to an improved understanding of the plasma chemistry and kinetics in these molecular discharges. For the investigation of plasmas, which are highly chemically active, various non-invasive diagnostic methods, like emission spectroscopy or low intensity laser absorption spectroscopy, can be applied. Tunable diode laser absorption spectroscopy (TDLAS) in the mid infrared spectral region between 3 and 20 µm is such a non-invasive technique for measuring number densities of stable molecules and radicals. TDLAS can also be used to determine neutral gas temperatures [1] and to investigate dissociation processes [2-5]. Due to their small laser line width (about 10^{-4} cm^{-1}) tunable lead-salt diode lasers are also used for high resolution spectroscopy purposes in the mid infrared region, e.g. of low molecular weight free radicals and molecular ions [6-9].

Plasmas containing hydrocarbon precursors are used in a variety of plasma enhanced chemical vapour deposition (PECVD) processes to deposit thin films with advantageous mechanical, electrical or optical properties. To open up new fields of application the deposition of coatings with a wide range of chemical and physical properties by varying plasma parameters is a challenging subject for plasma technology. In any case, the key to an improved understanding of plasma chemistry and kinetics in non-equilibrium plasmas containing hydrocarbons is the analysis of the fragmentation of the precursors and the monitoring of transient or stable plasma reaction products, in particular the measurement of their ground state concentrations. The decomposition of hydrocarbons in non-equilibrium plasmas containing reactive gases such as oxygen and nitrogen has been the subject of a variety of experimental and theoretical studies. Some recent examples concerned with nitrogen admixture shall be given here; a more comprehensive discussion related to oxygen admixtures can be found in [2,10-14].

The interest in hydrocarbon plasmas with admixtures of nitrogen arises from various applications including deposition of diamond layers [15-17] and of hydrogenated carbon nitride films [18-21], detoxification of combustion gases [22], plasma chemical conversion of methane to higher hydrocarbons [23] or studies of astronomical objects such as interstellar clouds and stellar atmospheres [24-25]. In addition, discharges in nitrogen and methane at low pressures are used to simulate the atmosphere of Saturn's moon Titan [26-29] and of Earth's primitive atmosphere [30], which had similar composition. Such types of plasma are also gaining importance in fusion physics [31], where the suppression of hydrocarbon deposition in methane / hydrogen plasmas was found to be possible by the injection of nitrogen under conditions similar to those prevailing in the divertor region of present fusion devices. In any case, they are of interest since atmospheric contaminations of hydrocarbon containing discharges, e.g. due to a leakage, result in gas mixtures with a certain nitrogen content.

Transient molecular species, in particular radicals, influence the properties of nearly all molecular plasmas, both in the laboratory and in nature. They are of special importance for several areas of reaction kinetics and chemistry. The study of the behaviour of radicals together with their associated stable products provides a very effective approach to understanding phenomena in molecular plasmas. Radicals containing carbon are of special interest for basic studies and for application in plasma technology. The methyl radical is generally accepted to be one of the most essential intermediates in hydrocarbon plasmas. For the *in situ* detection of the methyl radical only two approaches are suitable: optical methods and mass spectroscopy. In the latter case threshold ionisation mass spectrometry (TIMS) or photoionisation mass spectrometry (PIMS) have to be chosen for the detection of radicals [32-37].

Non-invasive optical methods for detecting the methyl radical are based on absorption spectroscopy either with 216 nm ultraviolet radiation [38,39] or in the mid infrared near 3100 cm^{-1} [40] or 606 cm^{-1} [41]. Beginning with Hirota and co-workers high sensitivity and resolution have been achieved using infrared lasers, in particular, tunable diode laser absorption spectroscopy (TDLAS) [41-42]. The infrared TDLAS technique has proven to be the most useful because it can also measure the concentrations of related species provided they are IR active. Outside of plasma diagnostics and high-resolution spectroscopy, this technique has been used successfully in the field of atmospheric trace gas monitoring and for exhaust gas monitoring of on-road vehicles, e.g. refs. 43-45.

TDLAS method also allows time resolved measurements with repetition rates exceeding 1 kHz. Recently, new compact and transportable TDLAS systems for plasma and gas phase process diagnostics and control have been developed. These use up to four diode lasers operating simultaneously and independently with a time resolution as short as ten microseconds [46,47]. For example, the time dependence of the conversion of methane to the methyl radical and three stable C-2 hydrocarbons was studied in surface wave plasmas using such a system [12].

The present work describes recent spectroscopic diagnostic studies of H_2-Ar-N_2 microwave plasmas containing methane or methanol in a planar microwave reactor (f= 2.45 GHz, p= 1.5 mbar). Over the past decade planar microwave plasma reactors have been intensively studied and used for plasma chemical applications. The interest in such discharges is that the microwave power is coupled rather efficiently to the plasma without any electrodes. Planar microwave plasma sources have been used for plasma chemical applications [48] diamond deposition, surface corrosion protection by deposition of organo-silicon compounds and surface cleaning [49-53]. In addition to studies oriented towards plasma technology, this type of plasma source has been used for investigation of several phenomena in molecular plasmas, in particular excitation and relaxation processes [54], and plasma chemistry and reaction kinetics [2], as well as basic molecular spectroscopy [55].

The main subjects of the work presented here are TDLAS studies of the plasma chemistry in non-thermal molecular plasmas using tunable diode lasers in the mid infrared region. Systematic investigations of the methyl radical and the stable molecules CH_4, CH_3OH, C_2H_2, C_2H_4, C_2H_6, C_2N_2, CH_2O, HCN and NH_3 have been performed in mixed H_2-Ar-N_2 microwave plasmas containing small percentages (0.9-7.2 %) of methane or methanol. The molecular concentrations in the plasma and the degree of dissociation of the added hydrocarbon precursor were monitored as the nitrogen concentration was varied while maintaining constant discharge pressure and total flow rate. In addition, the species concentrations in the plasma were measured under static conditions to investigate the nature of the plasma chemical processes under two limiting situations (while maintaining constant percentage of the precursor hydrocarbon gases in the mixture): first, in the relatively high gas flow regime (fc) and second, under static conditions (sc). The intermediate behaviour between these two extremes was analysed by time resolved TDLAS measurements.

The main experimental features of the present work can be summarized as follows:

(I)   Nine stable molecules, three of them nitrogen containing, and two radicals, in particular the methyl and the CN radical, have been measured.

(II)   The planar microwave plasma source has been combined with an optical multi pass cell ("White Cell") to improve the sensitivity of TDLAS measurements.

(III)   The degree of dissociation of the hydrocarbon precursors in discharges of gas mixtures containing nitrogen, hydrogen and hydrocarbons was determined.

(IV)   Intermediate as well as final products of the reaction kinetics were identified.

(V)   The temporal developments of ten species concentrations in the plasma have been measured for various gas mixtures and for both types of discharge conditions.

(VI)   18 $^{12}C^{14}N$ and 22 $^{13}C^{14}N$ line positions have been determined with high accuracy to derive molecular constants of both CN isotopes.

The spectroscopic studies of the CN free radical were done using TDLAS in CH_4-H_2-N_2-Ar plasmas. This radical has played a central role in the development of molecular spectroscopy. Its band systems are ubiquitous in flames and in discharges containing hydrogen and carbonaceous molecules and it has been detected in astronomical objects.

Chemical modelling of the methane plasma under static discharge conditions has been performed to predict the concentrations of those gaseous species which have not been detected so far. In total 145 reactions for 22 gaseous species were included in the model leading to relatively close agreement between experimental and calculated concentrations, and to an improved knowledge of the main chemical reaction pathways.

It should be emphasized that the present study is based on the experimental methodology used already in former investigations of H_2-Ar-O_2 plasmas containing methane or methanol [2]. The new aspects described in this thesis are fivefold:

(i)   a changed chemical system using nitrogen instead of oxygen in the plasmas,

(ii)   the monitoring of the transition between flowing and static conditions by time resolved TDLAS,

(iii)   an improved dynamic range of more than one order of magnitude for concentration measurements by a multiple pass optical arrangement,

(iv)   the estimation of fragmentation rates of the hydrocarbon precursors and of conversion rates to main reaction products, and

(v)   a more sophisticated plasma chemical modelling.

2   Diagnostics of Molecular Plasmas

Many different methods have been applied in the field of plasma diagnostics. This chapter focuses on diagnostic methods of low-temperature low-pressure non-equilibrium discharges in molecular gases [57]. These discharges are usually characterised by a small degree of ionisation. The main interest is the determination of the neutral gas composition and of the properties of the active plasma compounds.

Diagnostic methods can be divided into several groups. It is possible to distinguish between in-situ and non-in-situ diagnostics as well as to distinguish between invasive and non-invasive methods. The question is where to set the temporal or invasion limits. In this work, invasive and "not so invasive" (short form: non-invasive) methods are distinguished.

2.1   Invasive Methods

Invasive methods disturb the object of measurement, i.e. the discharge and its properties persistently. They extract particles or disturb the local plasma properties like the electron energy distribution function significantly.

Probe Measurements

The application of electric probes has a long tradition in the field of low-temperature plasma diagnostics. The method was developed in the beginning of the last century by Langmuir et. al. [58]. Since then, several extensions of the probe theory have been introduced [59-63]. In general, the current-voltage characteristic of the probe is studied. From that, basic parameters of the surrounding plasma, like the electron density, the electron temperature or the electron energy distribution function can be determined. Results of probe measurements in the microwave discharge used for the present work have recently been published [64].

Probes interact with the surrounding plasma in many ways. They change the electric field in the discharge due to the applied voltage on the probe. Their surface is sputtered or coated due to the interaction with the particles of the discharge. In addition, in chemically active discharges deposition of conducting or non-conducting layers on the probe surface can influence the measurement results. This diagnostic method is limited to the low pressures region.

Mass Spectrometry

Mass spectrometers are amongst the most common tools to be found in the field of plasma diagnostics [65-68]. The gas composition is investigated by analysing particles extracted from the discharge. The extracted neutral atoms or molecules are ionised in an ion source. The main part of a mass spectrometer is the ion separation system. Here the ions are separated according to their mass-to-charge ratio. The most commonly used type is the quadrupole invented by Paul and Steinwedel in 1953 [69]. A high frequency electrical field applied to four cylindrical rods interacts with the ion beam coming from the ion source in a normal direction. An ion detector quantifies the number of separated particles. The study of the kinetic energy of the ions requires an additional ion energy analyser. Time resolutions of several seconds down to some tenths of a second are possible. While the sensitivity is rather high ($\sim10^4$), the measurement of absolute concentrations is normally not possible.

Two types of mass spectrometric analyses of plasmas are possible, the particle flux analysis type ("plasma monitor") and the partial pressure analysis type [68]. In the flux regime, the particles fly directly without any collisions from the plasma into the analyser. A study of ions and ionised radicals including their kinetic energies is possible [35,70]. In the partial pressure analysis type the discharge chamber and the mass spectrometer are connected via a vacuum tube. A sufficient pressure reduction between the plasma and the ion source can be established, if necessary. With this analysis type, collisions of the particles with the wall and between the particles will appear before they reach the detector. Therefore, no ions or radicals can be detected and no energy analysis is possible. This type of analysis is widespread, e.g. for residual gas analysis or leak detection.

Various microwave discharges containing hydrocarbons have been investigated in the past using mass spectrometric methods [71,72]. However, for radicals like the methyl radical this type of diagnostic tool is not easy to apply. Radicals can appear as a real plasma product or as a fragmentation product of the ionisation in the ion source. This can be investigated by optimising the ionisation process. In this case, the ionisation energy of the ion source is varied. The threshold energies of ionisation and of dissociation are characteristics of every molecule. This method have been used for the determination of different radicals e.g. in rf methane plasmas [33,34]. Another option is the application of photo ionisation processes since the energy of ionising photons can be determined precisely [65,73]. This was used for measurements of the methyl radical, e.g. by Ando et. al [37].

Gas Chromatography

The method of gas chromatography also extracts gas from the volume under investigation [74]. Gas chromatography is based on the species distribution between two immiscible phases. Different species have different velocities in the separation column and reach the detector after different time intervals. It allows the qualitative and quantitative analysis of more complex molecular mixtures with a species sensitivity down to some ppm. The identification of the molecular species is established using other detection methods like mass spectrometry or absorption. The pressure difference between the discharge under investigation and the diagnostic system results in the need for calibration. No transient species like radicals or ions can be detected. The temporal resolution is in the range of minutes and therefore rather low. Nevertheless, gas chromatography is a standard method, e.g. in the pharmaceutical and chemical industry, and it is used for automatic controlling of complex gas mixtures.

2.2   Non-Invasive Methods

Non-invasive methods usually do not perturb the target system and its properties significantly. In general, these are spectroscopic diagnostic techniques, which use low-power or short-pulsed laser radiation or which observe the light emitted by the plasma. They provide information about atom, molecule and ion densities in their excited or ground states as well as about gas, rotational and vibrational temperatures. High temporal resolution can be achieved, while spatial resolution requires complementary optical techniques since spectroscopic methods provide line-of-sight information only [75].

Optical Emission Spectroscopy

Electromagnetic radiation can be emitted from low temperature plasmas over a wide spectral range from the ultraviolet to the infrared. The investigation of this light is called optical emission spectroscopy (OES). Modern OES spectrometers feature high spectral resolution, sensitivity near the single photon detection limit and time resolution in the nanosecond range. The emission spectroscopy is in principle an inverse problem. Usually the integrated intensities of emission lines in the line of sight are measured within a solid opening angle using a certain spectral resolution. If the discharge is homogenous over the solid angle, a local value of the intensity can be determined. Otherwise, theoretical inversion methods have to be applied. The spatially localized emission intensity of a spectral line determined this way allows the calculation of population densities of electronic (atoms) or rotation-vibrational levels (molecules), provided the specific transition probabilities are known and a proper calibration of the spectrometer has been done. The calculation of species concentrations in the ground state requires a theoretical model of excitation and de-excitation processes, appropriate cross-sections, transition probabilities etc. as well as knowledge about the electron energy distribution function [75]. Since these requirements are not easy to fulfil, in practise actinometrical methods are often applied [76,77]. In these methods measured intensities are referenced to a known fixed standard. Nevertheless optical emission spectroscopy has been successfully applied in the field of molecular plasma diagnostics e.g. for microwave plasmas used for diamond deposition processes [72,78,79].

Laser-Induced Fluorescence

In laser-induced fluorescence (LIF) measurements, the species under investigation is excited from a lower energy state to a higher energy level using a laser tuned to the specific transition wavelength. The fluorescence emitted in the de-excitation process of the higher energy level is measured spectroscopically resolved [80]. If collisional relaxation processes can be neglected and the plasma is optically thin, concentrations in the lower level may be deduced from the subsequent fluorescence signal. Since only a single energy level is selectively populated the fluorescence spectra is rather simple in comparison to the complex spectra emitted by the discharge itself. LIF can provide high spatial (millimetre) and temporal (nanosecond) resolutions. It has been applied to a wide range of species like the hydroxyl radical [81], the CH radical [82-85], the CN radical [86,87] or atomic hydrogen [88-90] in various types of discharges.

Coherent Anti-Stokes Raman Scattering

Light scattering methods like Coherent Anti-Stokes Raman Scattering (CARS) are of growing importance in diagnostics of low-temperature process plasmas. The CARS diagnostic method is based on the inelastic Raman-scattering effect of photons on molecules [80,91-94]. It can be used for the determination of densities, rovibrational population distributions or temperatures of molecular plasma species. In principle CARS is a parametric four-wave mixing process. Three input lasers, two of a special frequency (ω_p, $\omega_{p'}$) to pump the species of interest to a specific excitation level and one with the Stokes-frequency (ω_S) of the species, overlap in the medium under investigation. The frequency difference of the pump laser and the Stokes-laser has to be tuned to the Raman resonance frequency $(\omega_R = \omega_p - \omega_S)$ characteristic of the molecular species under investigation. Under these conditions, a scattered anti-Stokes laser-like beam is emitted from the interaction zone. Scattering with photons of the second coherent pump beam is used in addition to shift the anti-Stokes beam to shorter wavelength. The frequency of this coherent anti-Stokes beam is given by:

$$\omega_{as} = \omega_p - \omega_S + \omega_{p'}.\qquad(2.1)$$

Since the CARS-signal is usually fairly low, high power lasers are an essential part of the experimental set-up. The signal can be used to determine molecular properties like concentrations, rovibrational population distributions or temperatures. The spatial resolution is given by the size of the interaction zone of the crossing laser beams and therefore it can be as small as millimetres. Time resolution in the nanosecond region is possible. Special problems of this rather advanced diagnostic technique are caused by fluctuations of the interacting laser beams in direction, frequency or intensity. The CARS method has been applied to investigate major plasma species in pulsed microwave discharges containing nitrogen, especially the vibrational and rotational temperature of N_2 [95-99] as well as to measure CO [100] or various hydrocarbons in microwave excited plasmas [101]. Measurements in the reactive zone of a test reactor for hydrocarbon synthesis have been reported as well [102,103]. Related diagnostic methods like Resonance Enhanced CARS (RECARS) and Degenerate Four Wave Mixing (DFWM) can be used to measure molecular concentrations down to about 10^{12} cm^{-3}.

Absorption Spectroscopy

Absorption spectroscopic methods are well suited, in contrast to emission spectroscopy, to monitor concentrations not only in excited states but in the ground state as well. A wide variety of light sources, dispersive elements, detectors and data acquisition methods can be used [80]. The classic absorption spectroscopic experiment uses a continuous light source in combination with a narrow bandwidth frequency filter, in particular a spectrograph, and with a detector suited for the investigated spectral range. Continuously tunable, narrow-bandwidth, low-intensity light sources, like tunable lead salt diode lasers in the mid infrared region, can be used as well.

The probing light intensity has to be low in any case to avoid saturation effects. While the spectral position provides information for species identification, the line shape of an absorption line is connected to plasma parameters, e.g. external fields, gas temperature and pressure. The important advantage of the absorption spectroscopy over emission spectroscopy is that only relative intensities of the emitted light of the laser source and of the transmitted light after the absorbing medium have to be measured to determine absolute concentration values. No intensity calibration of the spectrometer is necessary. Like emission spectroscopy, absorption techniques can be applied almost over a wide spectral range from the vacuum ultraviolet to the far-infrared depending on the external light source used and the species of interest. In addition, the self-absorption of the plasma can also be investigated. In this case, the light from the plasma measured with a retro-reflector is compared with the signal without the retro-reflection or with the direct light signal from an identical discharge. Identical line profiles of the emitted and absorbed light are assured in this way and in particular important information about the optical thickness of the plasma can be determined.

Absorption spectroscopy relies on the decrease of the light intensity dI_v while passing through a medium for a distance dz. The absorption of light in general can be described by the Bouguer-Lambert-Law:

$$dI_v = -\alpha(v)\, I\, dz .$$ (2.2)

The absorption coefficient α is the fractional absorption dI_v/I per path length unit dz. The Bouguer-Lambert-law requires several conditions to be fulfilled:

a)   The light has to be monochromatic and parallel.

b)   The absorbing medium has to be homogeneous.

c)   Scattering and reflection on surfaces can be neglected.

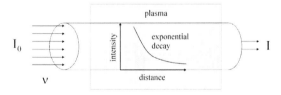

Figure 2.1: Radiation absorption in a transparent medium (Beer-Lambert-Law)

Beer showed 1852 that the absorption coefficient depends not only on the specific absorption line but on the concentration n of the absorbing species as well. Combining this fact with formula (2.2), one gets the Beer-Lambert-Absorption law:

$$dI_\nu = -\alpha(\nu)\, n\, I\, dz\,. \qquad (2.3)$$

If (I) the intensity of the emitted laser light is much higher than the radiation coming from the discharge and if (II) the absorption is continuous over the whole distance ($\alpha(\nu)$ = constant), formula (2.2) can be integrated giving with $I(z = 0) = I_0$ for the light intensity $I(z)$ at a position z:

$$I(z) = I_0 e^{-\alpha(\nu)\, n\, z}\,. \qquad (2.4)$$

The light intensity $I(z)$ decreases therefore exponentially as shown in Figure 2.1. The determination of the effective absorption length z and the assumption of a constant absorption coefficient over the complete distance z are two main problems of absorption spectroscopy in discharges. Small changes in the plasma, like inhomogeneities, anisotropies and temperature differences, can cause serious errors. Absorption spectroscopic methods provide in any case integrated intensities along the line of sight only. Recently tomographic methods have been applied in the field of absorption spectroscopy to resolve spatial distributions of plasma species concentrations [104].

The absorption coefficient $\alpha(\nu)$ can be written for a single line as the product of a line strength S of the specific transition and of the normalised line profile function f:

$$\alpha(\nu) = S\, f(\nu - \nu_0) \qquad (2.5)$$

with
$$\int_{line} f(\nu - \nu_0)\, d\nu = 1\,. \qquad (2.6)$$

While the value of the line strength can be taken from the literature, the influence of the line profile function has to be calculated separately. This is discussed in more detail for the direct absorption of a single line in the mid infrared region in chapter 3.2.

Cavity Ring-Down Spectroscopy

Cavity Ring-Down Spectroscopy (CRDS) is another high-sensitive laser absorption technique. It was invented in 1988 by O'Keefe and Deacon [105]. This method is based on the observation of the decay rate of a short laser pulse injected into an optical cavity formed by two highly reflective mirrors enclosing the plasma. Since the laser pulse is trapped in the cavity for many thousands of round trips, absorption path lengths in the kilometres range can easily be achieved [106]. For each reflection of the laser pulse, a small fraction leaks through the mirror, depending on the reflectivity of the mirrors forming the cavity.

Therefore, the light intensity detected beyond the mirror decays slowly to zero. Absorptions as low as 10^{-9} can be detected with an acceptable signal to noise ratio [107]. The temporal resolution is determined mainly by the detector response time and the ring-down time. It is in the order of some microseconds.

CRDS has numerous applications, e.g. in the field of atmospheric or plasma chemistry. It has been used for the detection of plasma species in etching discharges in fluorocarbon gases [108], of the CH radical in an Ar-C_2H_2-plasma [109] or of the CH_3 radical in HF-CVD reactors [110]. Recently CRDS was extended to the mid infrared region using tunable diode lasers [111].

Fourier Transform Infrared Spectroscopy
The main part of a Fourier Transform Infrared (FTIR) spectrometer is a Michelson interferometer. The light of a continuous infrared light source interferes with the light transmitted through the plasma. The intensity depends on the variable optical path difference between the mirrors in the two arms of the interferometer. While one mirror is fixed the other is moved continuously between identical path lengths (maximum signal) and a shifted path length of $\lambda/2$ (zero signal). If this mirror is moved continuously, an oscillating signal will be observed. On the detector a Fourier transformed signal of the incident radiation is detected. The inverse transform yields to a normal spectrum.

In contrast to dispersion techniques, the FTIR spectrometer records the whole spectrum simultaneously with spectral resolutions as high as 0.002 cm^{-1} and absorptions as low as 10^{-4}. The measurement speed is determined by the mirror speed and the wavelength range.

Tunable Diode Laser Absorption Spectroscopy
Tunable diode lasers are narrow-band low power light sources that can be used as a substitute for continuous light sources for absorption spectroscopy experiments. They are characterised by their high spectral intensity, by their small bandwidth, and by the capability of tuning the radiation over the absorption profile. TDLAS in the wavelength region between 3 and 20 µm has been applied in the last years e.g. in hydrocarbon containing plasmas. It is mainly used for measuring number densities of stable molecules or of radicals as well as for the determination of wave numbers of transitions with high accuracy. Typical spectral resolutions are in the range of 10^{-4} cm^{-1}, and absorbencies down to 10^{-4} or even lower can be observed. Therefore, depending on the line strength, a sensitivity of 10^{10} molecules cm^{-3} can be reached. Due to its high sensitivity and selectivity,

TDLAS is a promising technique for the investigation of molecular discharges. The detection of concentrations in the ground state allows the monitoring of the dissociation of precursor molecules and of the main plasma products. Spatial resolution can be achieved by reducing the laser beam diameter and moving the optical axis with an optical system. Time resolutions of several milliseconds [46] and even in the microsecond range [47] are possible.

Tunable diode laser absorption spectroscopy is the diagnostic method used for the measurements of this work because of its low detection limit. It was the method of choice since a wide variety of species in the plasma can be detected and absolute concentrations of different radicals and stable molecules can be measured. The special characteristics and methods of TDLAS are discussed in the following chapter.

3   Tunable Diode Laser Absorption Spectroscopy

Tunable diode laser absorption spectroscopy was introduced as a diagnostic method for high sensitivity gas monitoring in the field of environmental investigations. Many different atmospheric molecules, in particular those which are responsible for the green house effect or acid rain, have been investigated in detail using e.g. weather balloons or rockets. Most experimental set-ups include long-path multi-reflection cells for improving the sensitivity of the measurement [112].

Besides atmospheric compounds, most polyatomic and hetero-nuclear diatomic molecular substances in the gaseous state (gases and vapours) absorb infrared light in some parts of the infrared spectral range. Homo-nuclear diatomic molecules, such as O_2, N_2, H_2 or Cl_2, and monatomic gases, such as helium, argon or neon, do not have infrared bands, and therefore they cannot be measured in the infrared region.

Infrared light is absorbed at frequencies that match molecular vibrational or rotational frequencies. The spectral region between $500\ cm^{-1}$ and $3700\ cm^{-1}$ is called the "mid infrared" to distinguish it from the near infrared and the far infrared. Because of the different masses, bond strengths, bond angles, etc., each molecular species has its unique vibrational and rotational transitions. Few molecules have strong bands at higher wave numbers and only very weak overtones are detectable in this part of the spectrum. At lower frequencies there may be good strong rotational lines, but strong absorption by atmospheric water vapour makes their detection quite difficult [113].

The pattern of absorption is determined by the physical properties of the molecule such as the number and types of atoms, the bond angles and the bond strengths. The atomic mass and electronic structure of the atoms involved in molecular stretching and bending determine the rotational or vibrational frequency. This means that each spectrum differs from all others and may be considered as the characteristic molecular "fingerprint". Quantitative analysis can be performed by measuring the infrared absorbance at one of these frequencies. In the study of gases, the intensity of the infrared absorption depends on the total number of molecules in the path of the radiation. At a fixed total pressure, trace gas concentration and optical path length determine the absorption of a particular line. The line width of a specific absorption line depends on the gas temperature, on the total pressure or on other parameters. While electronic absorption in the UV region results in broad absorption bands, the absorption lines in the mid infrared are as small as 10^{-2} cm^{-1} for low pressures. To investigate such small absorption lines narrow-bandwidth light sources like tunable diode lasers are well suited.

3.1   Diode Lasers

Tunable lead salt diode lasers in the mid infrared region are narrow-band light sources well suited for absorption spectroscopy experiments. They have the advantage of sufficiently intensity, narrow bandwidths and continuous tunability over the absorption profile of a line. The structure of a diode laser in its simplest form, the homo structure diode laser, is illustrated in Figure 3.1. The PbSnSe or PbSSe semiconductor crystal with a dimension of several tenth of a millimetre has a p-n-junction and metal contacts that supply the current through the diode. Very divergent laser radiation is emitted due to the population inversion created in the active zone of the p-n-junction when a sufficient current flows. The front- and back-side of the crystal form the two facets of the laser resonator. The current limits for the laser radiation is a typical property of a diode laser. They depend on the diode laser type, operating temperature and operating mode (pulsed or cw) in the range from several 10's of mA to some Amperes. Figure 3.2 illustrates the real size of the diode laser. The semiconductor crystal itself is the small black spot on top of the white body.

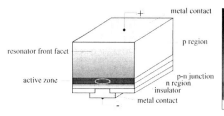

Figure 3.1: Homo structure lead salt diode laser

Figure 3.2: Size comparison of a mounted diode laser with a pin

While diode lasers for the visible or near-infrared region operate at room temperature or just below, lead salt diode lasers for the mid infrared region operate between 120 and 20 K. For this reason the tunable diode lasers used for the measurements for this thesis were cooled by a Helium closed cycle refrigerator system.

The wavelength range emitted by the diode laser is determined by the selected molar ratios of the crystal compounds. A rough tuning of the laser wavelength of ~4 cm^{-1} / K can be made using the operating temperature. This is due to a change of the gap between the conduction and the valence band of the semiconductor. A fine-tuning of 1 cm^{-1} / K is possible using the temperature change of the refractive index of the crystal. The continuous fine-tuning of a single laser mode can be achieved by changing the current applied to the diode. This causes a slight change of the temperature in the active zone and therefore a tuning of the emitted laser radiation of about 0.1 cm^{-1} / mA or less.

At a given temperature and current usually more than one laser mode of different wavelengths is emitted. Also a particular frequency can be emitted from different modes at different temperatures and currents. Typical examples for the mode behaviour of a lead salt diode laser in the mid infrared region are shown in Figure 3.3.

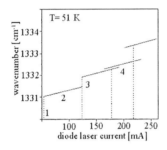

Figure 3.3: Example for the mode structure of a tunable diode laser
(1 – threshold, 2 – single mode, 3 – mode hop, 4 – multi mode)

3.2   Concentration Measurements by Direct Absorption Spectroscopy

Like already discussed in the paragraph devoted to absorption spectroscopy in chapter 2.2, the intensity I of the transmitted laser light depends besides the intensity I_0 emitted by the laser on the concentration of the absorbing species n, on the absorption length z and on the line specific absorption coefficient $\alpha(v)$. For the concentration of the species under investigation one gets therefore from equation (2.4):

$$n = \ln\left(\frac{I_0}{I}\right) \frac{1}{\alpha(v)\, z}. \qquad (3.1)$$

While the emitted and transmitted intensities and the absorption length can be measured, the absorption coefficient has to be calculated. It depends on the line strength and profile (equation (2.5)). The line parameters of many molecular species lines in the mid infrared region are listed in books [114,115] or other compilations like HITRAN [116,117] or GEISA [118-120]. While the absorption strength is constant, the line profile of a special line changes due to several different effects. For the absorption spectroscopic experiments applying tunable diode lasers in the mid infrared to low pressure gas discharges in principle the following aspects are of importance for the line shape:

1) Natural line width

2) Doppler broadening

3) Pressure broadening

4) Saturation broadening

An absorption line without further influences has the form of a Lorentz-profile. The natural line width results from the relaxation time of the state excited by the absorbed light. It is the smallest possible line width of a spectral line. Since it is usually in the order of 10^{-7} cm^{-1} only, it can be neglected in most cases.

Spectral lines are additionally broadened in almost all cases. Hence, the spectral intensity distribution of the absorbed light has to be taken into account for the calculation of species concentrations. Doppler-broadening is caused by thermal movements of the molecules in the gas, described by a Maxwellian velocity distribution. From the parallel and the anti-parallel velocity components of this movement in relation to the absorbed light, the Doppler-effect for the emission and absorption of light quanta is caused. The spectral line has the form of a Gaussian profile. The Doppler half width v_D (HWHM) of a spectral line is given by:

$$v_D = \frac{v_0}{c} \sqrt{2 k_B N_A \ln 2 \frac{T}{M}} . \qquad (3.2)$$

In reality the line profile is described in better accuracy by a Voigt-Profile, the convolution of Lorentz- and Gaussian profiles. At room or even higher temperatures Doppler broadening is dominant and the difference of the line profile fit using a Gaussian or a Voigt profile is negligible.

Spectral line widths vary not only with the gas temperature but with pressure as well. If the sample pressure is lowered, absorption lines get taller and narrower. This is caused by the decreasing number of molecules and collisions of the molecules. Pressure broadening can be neglected for measurements of gases in the low-pressure range (p< ca. 5 mbar) in comparison to the Doppler broadening effect.

The fourth effect, saturation broadening, is avoided by using lines with absorptions of 80 % in maximum, only.

Applying these simplifications one gets for the absorption coefficient $\alpha(\nu)$:

$$\alpha(\nu_0) = \frac{S(\nu_0)}{\nu_D}\sqrt{\frac{\ln 2}{\pi}} \ . \tag{3.3}$$

The concentration can be determined for measurable absorptions (> 1 %) of gases in the low-pressure range therefore by:

$$n = \ln\left(\frac{I_0}{I}\right)\frac{1}{\sqrt{\ln 2/\pi}}\frac{\nu_D}{S(\nu_0)\cdot z} \ . \tag{3.4}$$

3.3   Derivative Spectroscopy

Direct absorption spectroscopy has its sensitivity limits for very small absorptions (< 5 %). The sensitivity can be increased by an order of magnitude applying the method of derivative spectroscopy. Like in the case of direct spectroscopy, the frequency of the diode laser is tuned by a current ramp. In addition, a small modulation of the laser frequency is overlapped due to a corresponding modulation of the applied current. The modulation has to be small in comparison to the width of the absorption line. The intensity change due to the modulation is detected using phase sensitive lock-in amplifiers.

The laser frequency ω_L is overlapped by a sinus modulation of frequency Ω and amplitude a

$$\omega_L = \omega_0 + a\sin\Omega t \ . \tag{3.5}$$

If now the phase sensitive amplifier is locked to the harmonic $i\Omega$ of the modulation frequency Ω, in principle the i-th derivation of the specific absorption $\alpha(\nu_0)$ n is measured. Therefore lines with rather small absorptions can be measured, even with a high background noise level. This method is discussed in more detail in [80]. It was used in particular for the spectroscopic investigations described in chapter 5.6.

3.4   Line Identification and Spectral Calibration

The identification of lines and the measurement of their absolute positions were carried out using well-documented reference gas spectra and an etalon of known free spectral range (FSR) for interpolation [114-120]. An important precondition for the accuracy of line identification is the quality and volume of the database of spectral line positions in use. In the last decades significant progress had been achieved e.g. in the HITRAN database [116,117] or the GEISA database [118-120]. The superior accuracy of spectral standards allows the calibration of spectra recorded near the Doppler limit for the resolution.

In the present work, in the case of enough absorption (> 2 %) the method of direct absorption spectroscopy could be used to measure concentrations. For lines with smaller absorptions, e.g. those of the CN radical, the lines were recorded using diode laser current modulation and phase sensitive detection at the second harmonic of the modulation frequency (second derivative method). As an example an absorption line of the CN radical in a CH_4-N_2-Ar-discharge chamber, of OCS as known reference and an etalon recorded by a three-channel spectrometer (see chapter 5.6.1 and Figure 5.38) using phase-sensitive detection with lock-in amplifiers is shown in Figure 3.4.

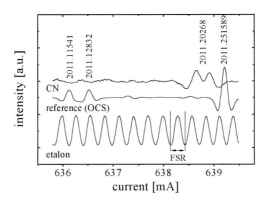

Figure 3.4: Calibration of the spectral position of a CN line using a reference gas (OCS) and an etalon with a known fringe spacing range (FSR = 0.01 cm^{-1})

Spectral positions of stable and commercial available gases can be identified using special gas cells filled with them as reference sources. In this case the identification can be carried out directly. If the molecular gas of interest is not available or unstable the spectral position is identified with the aid of one or more reference gases which have absorption lines at the same wave number position (Figure 3.4).

As a first attempt for the approximate determination of the spectral position one uses the wave length given by the calibration curve of the monochromator used as mode filter. If the calibration of the monochromator was performed carefully using the orders of a HeNe laser the diode laser wave number can be determined with an uncertainty of 1 cm^{-1}. The accuracy of wave number determination is enhanced using reference gases and etalon. For the wave number given by the monochromator a spectra of the molecule of interest can be simulated by software routines using spectroscopic databases like HITRAN or GEISA. This unique pattern of absorption lines is usually compared with the investigated absorption pattern. The wave number difference between measured lines is interpolated using an etalon. This Fabry-Perot etalon for the infrared range is a solid Germanium crystal with known fringe spacing range. The relative intensities of calculated and observed absorption lines should be similar.

The identification of molecular spectra can be complicated by overlapping absorption lines of atmospheric gases, in particular of carbon dioxide and water. These background absorptions cover wide parts of the mid infrared range. The atmospheric absorption lines are recognizable in the spectra by their high-pressure broadened line width. Additional water concentrations in contaminated reference gas cells cause a small low-pressure peak on top of the high-pressure background absorption line. This typical pattern can be used for the determination of wave number positions as well.

4   Plasma Chemistry

Plasma technology, and in particular, the field of plasma chemistry processing influence and improve the quality and performance of various industrial products and processes in many different ways [121,224]. Although it is already exploited in several areas, like in light source production and surface finishing, it is still a young and promising technology. Its full potential has not been realized so far. Examples of newer applications include plasma reactors such as exhaust purifiers for environment protection, functional coating of architectural glass, mercury-free lamps, plasma treatment of materials for the food industry and the production of nano-materials [122]. While most of these applications are surface-based the plasma itself is determined by its production and composition. Thus, the plasma and chemistry processes determine the interaction processes with the wall. Surface treatments, including activation and fine cleaning, functionalization and coating, and the plasma boundary layer processes, are all based on chemical reactions in the discharge.

Hence, plasma chemistry can be applied for other purposes than light sources or surface modifications. It is also in use to synthesize special gaseous compounds. Electronic collision processes cause the production of free radicals in the discharge. By specific changes of external and in series of internal discharge parameters, it is possible to specifically promote certain reactions and to suppress others. For instance, control of the electron energy distribution function can be achieved by changing discharge parameters of pulsed dc plasmas like the duty cycle or the current density. Another motivation for applying plasma chemistry is the possibility to develop alternatives to catalytic processes. In this field, expensive and sensitive conventional chemical catalysts are to be replaced by plasma chemical means [224].

In general, plasma chemical conversion processes can be distinguished by the state of the equilibrium in the plasma as thermal and non-thermal. While in thermal plasmas the chemical conversions are caused by a similar kinetic temperature of all plasma compounds, in particular, of the electrons and ions, in non-thermal discharges the energy for the chemical conversion is transferred by electronic collisions only. Non-thermal gas discharges are characterized by their different energy distribution functions and mean energies, in particular, of their heavy compounds (ions and neutrals) and of the light electrons. Typically, the gas temperature of neutral particles is here some hundred Kelvin while the mean energy of the electron component is equivalent to a kinetic temperature of several

10^4 Kelvin. By pulsing or modulating of the discharge with high frequency, as in the case of microwave plasmas, it is possible to establish low ion energies while the electron component is the primary energy carrier. This is advantageous since such discharges are well suited for non-destructive surface processing.

To improve the efficiency and quality of the surface modification a detailed knowledge not only of the surface-wall interaction but also of the kinetics in the plasma zone itself is of particular importance to step up from an empiric attempt to a specific process optimisation. In particular, the properties of such non-thermal discharges like in microwave plasma sources enable the independent manipulating of the energy distributions of the different plasma species. By choosing a suitable plasma reactor configuration, gas pressure, gas mixture and power incoupling it is possible to separately control the temperature of, on the one hand, ions and neutral particles, and, on the other, that of electrons. By doing this, special reaction paths can be favoured while others can be suppressed. But, it is mainly the energy distribution of the electrons, which is important for initiating chemical reactions in the plasma zone of such non-thermal plasmas. Its control by changing macroscopic discharge parameters is therefore the key to an improved surface modification process. In their special properties lies the great advantage of low-temperature nonequilibrium plasmas for plasma technological processing as used in the present work.

In particular, the volume chemistry of low-pressure, non-equilibrium, molecular microwave discharges is of growing interest since these plasmas are characterized by a high number of chemical active species. Plasmas containing hydrocarbon precursors are used in a variety of plasma enhanced chemical vapour deposition (PECVD) processes to deposit thin films with advantageous mechanical, electrical or optical properties. To open up new fields of application, the deposition of coatings with a wide range of chemical and physical properties by varying plasma parameters is a challenging subject for plasma technology. In any case, the analysis of the fragmentation of the precursor molecules and the monitoring of their conversion chemistry is of great interest for the deposition of layers as well as for the use of plasmas for chemical conversion purposes.

An extensive overview of the technological potential of molecular low-temperature non-equilibrium plasmas is given in ref. 224.

Plasma chemistry conversion processes are not only important for producing deposition and polymerisation precursors. Discharges are used as well for exhaust gas cleaning purposes like the NO_x reduction and for the syntheses of different hydrocarbons like the acetylene syntheses from natural gas. The plasma based syntheses of chemical substances should allow an optimised production process with a smaller number of technological process steps.

The investigation and appropriate use of plasma chemistry processes and the understanding of its kinetics require detailed knowledge of the main plasma parameters, which can be obtained by appropriate diagnostic techniques. The online monitoring of transient or stable plasma reaction products in chemical reactors, in particular, the measurement of their ground state concentrations is the key to an improved understanding of these molecular discharges. For the investigation of plasmas, which are highly chemically active, various non-invasive diagnostic methods, like emission spectroscopy or low intensity laser absorption spectroscopy, can be applied. Tunable diode laser absorption spectroscopy (TDLAS) in the mid infrared spectral region between 3 and 20 µm is such a non-invasive technique for measuring number densities of stable molecules and radicals in gaseous media.

5   Investigations in Planar Microwave Discharges

5.1   Introduction

Plasmas can be generated in many different ways [123]. The most common method to produce a low temperature discharge in a neutral gas is by applying an electric field to the gas. Any volume of a neutral gas always contains free charge carriers formed e.g. due to the cosmic or natural radioactive radiation. These electrons and ions are accelerated by the electric field and produce new charged particles by collisions with neutrals.

Hence, low temperature plasmas can usually be classified by the temporal behaviour of the sustaining electric field. Besides dc discharges in particular ac plasmas, i.e. discharges excited by oscillating or pulsed fields, are used for technological purposes. A wide range of frequencies and electrode configurations is possible for discharges excited and sustained by high-frequency electromagnetic fields. RF discharges usually operate in the frequency range $f = 1...100\,$MHz. The range of microwave frequencies starts at about $300\,$MHz. All frequencies in this part of the electromagnetic spectrum have to be legally permitted. At these special frequencies the wavelength of the electro magnetic radiation is typically of order of the reactor dimension or smaller. This is an important property since it can cause inhomogeneities in the plasma. Microwave discharges usually operate at a frequency of $2.45\,$GHz. For this frequency, the wavelength is $12.24\,$cm. All experiments described in this works were done, if not mentioned different, with this frequency. The skin depth of the microwave into the plasma is limited to a few centimetres under these conditions.

Plasmas excited and sustained by microwave fields have several advantages for industrial proposes. They can be driven electrodeless. This allows very clean conditions in the discharge chamber and wide plasma extensions. In addition, microwave discharges in molecular gases dissociate the precursors in a high percentage. This leads typically to a high chemical reactivity due to the large number of transient and stable chemical active neutral species present in the discharge. Thus they are of interest for plasma chemical investigations as well as for spectroscopic investigations of diverse molecular species.

5.2   Power Transfer in Microwave Discharges

The main characteristic of the plasma phase is the existence of free charge carriers, i.e. of electrons and ions. If the ionisation degree is smaller than one, neutral particles, i.e. atoms, molecules, radicals and exited species, exist in the plasma too. For discharges excited and sustained by external electrical fields usually, due to the respective masses only the electrons can follow the electrical field and have therefore significant energies.

For this reason it is sufficient to describe the interaction of the electric field with the plasma by a simplified approach, that only takes the influence of the harmonic electric field and of elastic collisions of electrons with the neutral background gas into account [124]. Since the thermal motion of the electrons can be neglected compared to the motion in the field the equation of motion for a single electron is

$$m_e \frac{d\vec{v}_t}{dt} = -e\vec{E}(t) - m_e v_c \vec{v}_t . \tag{5.1}$$

Here the right side is the sum of a term describing the force of the electric field on the electron and a second term that represents the interaction of the electron with the heavy gas particles of the background due to collisions with a frequency v_c. Using the complex form of the velocity vector $\vec{v}_t = \mathrm{Re}(\vec{v}_t \exp(i\omega t))$ and time integration gives a relation for the complex amplitude of the electron velocity

$$\vec{v}_t = -\frac{e\vec{E}}{m_e} \frac{1}{v_c + i\omega} . \tag{5.2}$$

The transmission of an electromagnetic field in a medium is described in general by Maxwell's equations. Using this one gets the material constants of the plasma. The electron current density is for an electron density n_e

$$\vec{j} = -e n_e \vec{v}_t = \sigma \, \vec{E} . \tag{5.3}$$

The permittivity σ and the permeability ε are therefore

$$\sigma = \frac{n_e e^2}{m_e} \frac{1}{v_c + i\omega} , \tag{5.4}$$

$$\varepsilon = \varepsilon_r + i\varepsilon_i = 1 - i\frac{\sigma}{\omega\varepsilon_0} = 1 - \frac{\omega_{pe}}{\omega} \frac{1}{1 - i\frac{v_c}{\omega}} , \tag{5.5}$$

$$\text{with } \omega_{pe} = \sqrt{\frac{e_0^2 n_e}{\varepsilon_0 m_e}} \ . \tag{5.6}$$

ω_{pe} is defined as the electron plasma frequency. This characteristic value of plasma describes the oscillation of the electrons in relation to the relative fixed ions and is important for the understanding of the appearance of local charge fluctuations. Therefore, although the material constants only depend on internal parameters of the plasma they describe macroscopic properties of the plasma as well.

Using the relation for the permeability ε one can discuss the interaction of the plasma with the external electromagnetic field. Depending on the three characteristic values v_c, ω and ω_{pe} three cases can be distinguished:

(I) $\omega \ll v_c$ σ is real and the plasma behaves therefore like a conductive medium. Power can be transferred from the field to the plasma.

(II) $v_c < \omega < \omega_{pe}$ This case is a transition region. The plasma is still conductive. The power transfer nearby the plasma frequency is very efficient.

(III) $\omega > \omega_{pe}$ The plasma is a dielectric medium. No power can be absorbed.

In similarity with equation (5.6), where the electron density and the electron plasma frequency are linked, one can define a relation between the excitation frequency of the external field and the electron density:

$$n_c = \frac{m_e \varepsilon_0}{e^2} \omega^2 \ . \tag{5.7}$$

This so-called cut-off density is the critical value that specifies the switch of the plasma from a conductive to a dielectric medium. Since power incoupling is most efficient in the transition region near the plasma frequency, one can determine a minimum electron density for a special plasma. For e.g. 2.45 GHz the electron density has to be in the magnitude of 10^{11} cm^{-3} or higher, since $n_c \approx 7 \times 10^{10}$ cm^{-3}.

The importance of taking collisions of electrons with the neutral background particles into account shows a discussion of the collision free case ($v_c = 0$). The permeability (equation (5.5)) is now a real number ($\varepsilon = \varepsilon_r$) and the plasma is a medium without losses. The electrons oscillate ordered in the direction of the external field only. On average, no energy is transferred from the field to the free electrons and from the electrons to the heavier particles. The plasma would behave like an entirely passive medium in this case.

Elastic collisions of the electrons with the background gas introduce now a disorder into the motion of the electrons. Power is transferred from the field to the electrons and finally from the electrons to the heavier particles. The average kinetic energy an electron can pick up in one period in the field is

$$\bar{u} = \frac{e^2 \hat{E}^2}{2m_e \left(v_c^2 + \omega^2\right)} \qquad (5.8)$$

$$\text{with } \hat{E}^2 = \frac{EE^*}{2}.$$

E^* denotes the conjugate complex. This picked up energy is now much smaller than the energy necessary for ionisation processes induced by collisions. For microwave discharges e.g. with typical values of $\hat{E} \approx 50$ Vcm^{-1} and $v_c \approx 10^{10}$ s^{-1} the energy gain per period is about $\bar{u} = 1.6 \times 10^{-2}$ eV, which is more than two orders of magnitude lower than typical ionisation energies. In addition, due to the big mass difference not much energy can be transferred from the electron to the heavy particle with elastic collisions. However, the collision disorders the motion of the electron. It can pick up again energy from the external oscillating field. The mean transferred power per electron is

$$\Theta_A = -e\overline{\vec{E}(t)\vec{v}_t(t)} = \frac{e^2}{m_e} \frac{v_c}{v_c^2 + \omega^2} \hat{E}^2 = 2v_c \bar{u}. \qquad (5.9)$$

After several periods, it has enough energy to overcome excitation or ionisation limits. In this case, it can transfer its energy almost completely to the heavy particle by an inelastic collision. After this, the electron will be accelerated in the external field again.

So far, only the interaction of a single electron with the external field and with the background gas particles was discussed. A complete description of all electrons in a plasma makes a statistical approach necessary that leads to the definition of the electron energy distribution function (EEDF). Accordingly, few electrons have very high energies in the magnitude of excitation or ionisation energies of the atoms or molecules in the plasma. Therefore, they play an important role in the excitation and sustention of the plasma.

In a stationery discharge not only the energy input and loss is balanced but the number of charged particles too. I.e. the formation of electron ion pairs is equal to the loss due to diffusion or recombination in the volume or on the wall.

The strong dependence of the power transfer into the plasma on the frequency of the external field was already discussed. Besides the collision frequency of the electrons with the neutral background gas and the electron plasma frequency influence the plasma properties too. E.g. for similar power densities higher electron densities are found in microwave discharges than for lower frequency rf discharges [125]. With increasing discharge frequency the per-electron absorbed power, which is necessary to sustain the discharge, decreases. Besides, the average electron energy and the shape of the EEDF change with frequency [126]. The EEDF has more electrons with higher energies while the average electron energy decreases. Several papers have been published about the dependence of the EEDF on the discharge frequency of quasi-stationary plasma [124,125,127,128]. For low frequencies, the electrons can follow the oscillations of the electric field. The per-electron transferred power (equation (5.9)) depends on the electron energy and on the frequency of the external field. If the collision frequency and the field frequency are equal, the power transfer into the discharge is maximal. The collision frequency depends in general on the electron energy and on the gas. So it is an important factor for the shape of the EEDF too. If electron-electron collisions and collision of electrons with excited species can be neglected, e.g. due to the small number of excited species or the in low-pressure microwave discharges typical small degree of ionisation ($\sim 10^{-5}$), the EEDF only depends on the ratios $\hat{E}/N$ and ω/N but not on the electron density. N is the neutral species concentration.

With increasing external frequencies, even the electrons can no longer follow the oscillations of the field and an average value of losses appears. Therefore, the EEDF becomes time-independent and is determined by the $\hat{E}/N$ ratio. This value is very different for various gases or gas mixtures.

The shape of the EEDF determined by the v_c/ω ratio. If the collision frequency is much bigger than the excitation frequency, the EEDF changes quasi-stationary in the alternating field. The electrons behave in this case like in a dc-field of the positive column. If the frequency ratio becomes smaller than one, the EEDF changes to a Maxwellian shape to smaller values of the ratio.

5.3   The Planar Microwave Discharge

Microwave plasma sources can be constructed in many different ways. The main feature of microwave discharges is that they are excited and sustained electrodeless. Examples for reactor types are electron cyclotron resonance, surface wave and resonator discharges. The different discharge types are discussed in more detail in [129].

As mentioned in chapter 5.2 a characterizing effect of microwave discharges is the limited skin depth, which can cause plasma inhomogeneities. This has to be taken into account in the process of constructing a microwave discharge reactor. In addition, multi-mode fields can be excited. This can result in unstable switching between different discharge modes and therefore increases plasma inhomogeneities. To minimize plasma inhomogeneities usually non-resonant microwave structures with travelling wavefronts without any magnetic field are preferred. The microwave plasma source used in this work was constructed this way [2,48]. The special waveguide construction of this source allowed the production of large area planar microwave discharges.

5.3.1   Fundamentals

The phenomena in the planar microwave plasma source can be understood with a discussion of the propagation of an electromagnetic wave in a hollow waveguide at microwave frequencies. This chapter is not focussing on the theory of wave guiding. Only the special case of the propagation of the fundamental magnetic wave, the H_{10}-mode, in a rectangular waveguide is discussed here.

The microwaves are guided from the microwave generator to the discharge volume by hollow waveguides. These waveguides have a rectangular shape with a side ratio of $a : b = 2 : 1$. A schematic view of the waveguide with characteristic patterns of the wall currents is shown in Figure 5.1.

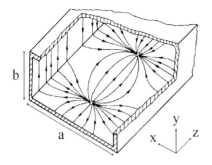

Figure 5.1: Wave guide coordinates and wall current distribution of the H_{10}-mode [130]

The current flows on the small sides (b) of the waveguide perpendicular to the direction of propagation of the microwave. On the wide side (a), it shows a complex pattern. An important characteristic for the use of the waveguide is the field structure of the propagating electric and magnetic fields. This is shown for the H_{10}-mode in Figure 5.2. As one can see, all field parameters, and so the electric field $\vec{E}$, have no components in the direction of the y-axis. This is an important feature of this type of waveguide for the H_{10}-mode and leads to its use in microwave plasma sources.

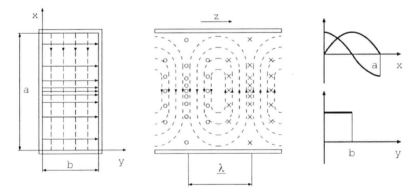

Figure 5.2: Field distribution of the H_{10}-mode in a rectangular wave guide (solid lines: electric field lines; dashed lines: magnetic field lines) [131]

For the realization of homogeneous microwave plasmas, the microwave is coupled gradually into a second rectangular waveguide. This is illustrated in Figure 5.3. The interface waveguide and the primary waveguide from the microwave generator form a T-shape-configuration. The graduate incoupling into the interface waveguide is done inductively by discrete adjustable metallic coupling elements with a distance of $\lambda/4$ (Figure

5.4). Since all ends of the two waveguides are terminated by matching microwave absorbers (see Figure 5.5) no resonant waves can appear. The microwave energy from the generator is transferred over the complete length to the interface waveguide. The main benefit of using an interface waveguide is that the discrete power incoupling of the coupling elements and the residual ripple in plasma homogeneity is suppressed. The lower metallic wall of the interface waveguide is replaced by quartz windows, which are transparent for the microwave field and which separate the waveguide (under atmospheric pressure) from the discharge volume (at low pressure). Under specific conditions, therefore a microwave discharge can be excited in the chamber. The plasma forms then the fourth conductive wall of the interface waveguide. A large area planar microwave plasma under the quartz windows can be produced by adjusting the coupling elements. The width of the waveguide limits the size of the microwave window to about 5 cm, but several parallel waveguides, which can have lengths of some meters in maximum, are possible to be combined [133]. Problematic for these configurations is the proper adjusting of the coupling elements.

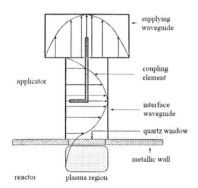

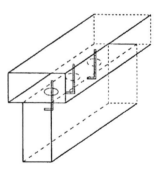

Figure 5.3: Cross-cut of the planar microwave applicator [48,132]

Figure 5.4: Microwave applicator with several coupling elements [48,132]

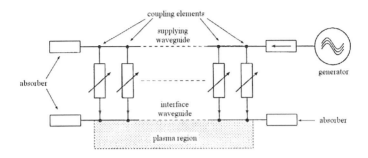

Figure 5.5: Technical scheme of the planar microwave applicator [48,132]

5.3.2 Properties

The external electrical field determines the properties of any type of electrical discharges. The ideal distribution of the electrical field strength $\vec{E}$ in a planar microwave plasma source with a single interface waveguide is shown in Figure 5.6. The electrical field is here assumed to be similar to that in the waveguide. The field strength in y-direction should be constant; no x-component should appear. At the transition from the quartz window to the metallic walls some wall currents flow into waveguide material. The strength of the electric field is distributed like in a plane-parallel capacitor. It decays into the plasma exponentially. Measurements in argon plasmas showed deviations from the theoretical field distribution, caused by non-ideal adjusting and orientation of the coupling elements [133].

Figure 5.6: Schematic of the ideal field strength distribution in the plasma zone of the planar microwave applicator

The skin depth of the microwave depends mainly on the permittivity σ and therefore on the electron density and frequency of the elastic electron collisions with the background gas particles. Using a microwave of 2.45 GHz, a spatial constant electron density of $n_e = 10^{11}$ cm^{-3} and an electron collision frequency of $\nu_c = 10^{10}$ s^{-1} the skin depth is calculated to be 19 mm [132].

A typical property of planar microwave discharges is the big number of excited neutral particles in the plasma. Therefore, their high chemical reactivity is a typical property. The plasma excited underneath the microwave window can be described by a simplified two layer model consisting of a preferential physically active plasma layer near the microwave windows followed by a preferentially chemically active decaying region [78]. In [52] the maximum density of activated neutral species achievable was estimated to be 4...6 orders of magnitude higher than the charged particle density, which was about 10^{11} cm^{-3} in maximum. The degree of dissociation is under these circumstances in the order of $10^{-5}...10^{-4}$. Nevertheless the electrons are responsible for the energy transfer from the external electrical field to the heavier particles, like discussed above. Plasmas in the pressure range of $10^{-2}...10$ mbar can be generated in the microwave plasma source described here since this device utilizes collisional power transfer processes between the microwave field accelerated electrons and neutral particles without magnetic fields. The electrical field strength in the plasma nearby the microwave windows is typically of the order of 10 V cm^{-1}. The power density varied between 2 and 20 W cm^{-2}. Typical values of E/N are in the order of 10^{16} V cm^2. Electron densities of 10^{12} cm^{-3} should be possible to achieve in this plasma source, but measurements showed lower values in the range of $10^9...10^{11}$ cm^{-3} [78,132]. For density values smaller than $n_c \approx 7 \times 10^{10}$ cm^{-3} the gas is transparent for the microwaves.

Due to the small charge carrier concentrations no Coulomb interaction should appear, the plasma is collision dominated. The electron temperature near the microwave windows is about 1.5 eV and decays into the discharge volume [79]. In average it is 1 eV in the active plasma zone. The gas flow in the chamber can be described using a plug flow model [52]. The gas flow speed is therefore in the order of 1 m s^{-1}. This is much smaller than the thermal movement of the particles. For this reason diffusion processes dominate the transport in the discharge volume.

5.4   Experimental System

5.4.1   Discharge Reactor System

The experimental set-up of the planar microwave plasma reactor and the tunable diode laser (TDL) system is shown in Figure 5.7 - Figure 5.10. It was changed to adapt it to the actual measurement requirements. Details of the reactor and spectrometer can be found elsewhere as well, e.g. in [2,133]. The discharge configuration in planar microwave plasmas has the advantage of being well-suited for end-on spectroscopic observations because a considerable homogeneity can be achieved over relatively long plasma path lengths. The dimensions of the vessel used here were $150 \times 21 \times 15$ cm^3. As in the study described in ref. 2 the measurements were performed at a microwave frequency of 2.45 GHz and power of 1.5 kW in an unpulsed plasma regime, corresponding to a power flux in the TDLAS observation plane of about 10 W cm^{-2}. It was proofed under different conditions that the plasma chemistry does not change significantly by changing from an unpulsed to a pulsed microwave plasma.

The gas flows were measured on mass flow controllers and then the gases mixed before entering the reactor. They were pumped out via a port in the reactor wall diagonal to the gas inlet. The gas mixture supplying the reactor consisted, if not mentioned different, of x sccm H_2 + 60 sccm Ar + y sccm N_2 + z sccm CH_4 or CH_3OH. The total flow rate and pressure were kept constant at 555 sccm and 1.5 mbar respectively. The pumping speed was adjusted with a butterfly valve to maintain a constant pressure in the plasma under flowing conditions. The pressure was measured using a capacitance pressure gauge. For both methane and methanol containing gas mixtures four different flow rates (z) were selected: 5, 15, 25 and 40 sccm corresponding to 0.9, 2.7, 4.5 and 7.2 % of the total gas flow rate and mixture, respectively. When nitrogen was added the proportion of hydrogen was reduced correspondingly to maintain a constant flow rate.

planar microwave plasma reactor

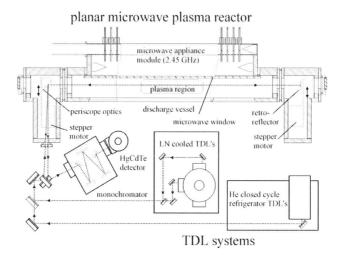

TDL systems

Figure 5.7: Experimental arrangement of the planar microwave plasma reactor (side view) with periscope optics and two tunable diode laser infrared sources. The laser beam path is indicated by dotted lines.

Figure 5.8: Photo of the planar microwave plasma reactor with periscope optics mounted

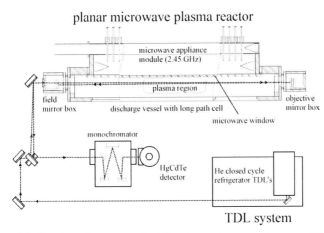

planar microwave plasma reactor

Figure 5.9: Experimental arrangement of the planar microwave plasma reactor (side view) with White cell multiple pass optical arrangement and tunable diode laser infrared source. The laser beam path is indicated by dotted lines.

Figure 5.10: Photo of the planar microwave plasma reactor with White cell optics mounted

The most of the measurements were performed with the infrared beam passing twice through the reactor volume, using periscope optics and a retro-reflector arrangement [2] (Figure 5.7 and Figure 5.8). For monitoring of CH_2O, C_2N_2 and of CN as well as for time resolved concentration measurements of the methyl radical multiple pass optics were used to achieve higher sensitivity (Figure 5.9 and Figure 5.10). With the White cell installed 24 passes were realized, leading to an optical length inside the reactor of about 36 m in maximum [135]. The TDL setup is described in more detail in chapter 5.4.2.

The plasma length varied between 12 and 20 cm depending on the discharge conditions and on the microwave incoupling. For the purpose of concentration measurements, the unstable species, like the methyl radical, were assumed to exist only in the visible active plasma region of the reactor. On the other hand, stable molecules were assumed to diffuse into the side ports, in particular into the boxes of the periscope optics and the White Cell mirrors, through which the diode laser beam passed as well. For concentration measurements of stable species, therefore, the path length was taken to be the complete optical path inside the discharge reactor. In order to determine the translational temperature of the plasma an experiment using optical emission spectroscopy was carried out [136-139]. This investigation gave a neutral gas temperature in the discharge of 1000 ± 100 K in the case of plasmas having high hydrogen content. Otherwise, a temperature of 300 K was used for the calculation of the concentration of the stable species outside the plasma region itself. The lines used to deduce the concentrations of all stable molecules and of methyl were found to be relatively insensitive to temperature.

In order to eliminate the influence of diffusion processes inside the plasma vessel and side ports an empirically determined method, already described in [2], was applied for the concentration measurement of stable species under static conditions and for the time resolved concentration measurements. Briefly, the plasma reactor was first thoroughly evacuated and then filled with the flowing gas mixture of interest. Once the pressure had stabilised the microwave generator was switched on and after 40 seconds the concentration measurements under flowing conditions were made. With the plasma on, the gas flow and the vacuum pump were stopped simultaneously to isolate the reactor while maintaining constant pressure as far as possible (variations were less than 15 %). At this point the time resolved measurements were started. After another 60 seconds concentration measurements in the closed reactor were made. The temperature dependence of the line strengths of the absorption lines was proofed to be negligible by taking another spectra after switching off the plasma.

5.4.2   Tunable Diode Laser Arrangement

The narrowband infrared emission of lead-salt diode lasers was used to monitor the infrared absorption features of the target species. The arrangement of the instrument set-up on an optical table is illustrated in Figure 5.7 - Figure 5.10 too. Four diode lasers could be mounted in a cold station (Model L5731, Laser Analytics Inc.). The temperature of each laser is controlled at milli-Kelvin precision in the range between 20 K and 110 K.

The divergent laser emission is collimated by an off-axis parabolic mirror (OAP). A HeNe laser and two diaphragms (not shown in the figures) are used for the alignment of the optical arrangement and of the multi pass cell.

The infrared diode laser beam from the TDL source assembly entered the plasma chamber via a KBr window. The periscope optic mirror system, which was combined with a retro reflector, was for an improved measurement sensitivity later replaced by a White Cell multiple pass setup [140]. The periscope optics were used for spatial resolved investigations depending on the distance to the microwave incoupling window. The distance of the infrared beams below the microwave window was set to about 2-3 cm. Otherwise, the multiple pass optics provide a higher number of passes up to 24 in maximum. To accomplish the image relay two objective mirrors and one field mirror with spherical optics were used. One of the advantages of the White cell design is the possibility to establish different path lengths. The light alternatively strikes each objective mirror, so that a light path trough the cell makes a multiple of four passes in which it hits each of the two objective mirrors once and the field mirror twice. The laser beam path in a White cell is illustrated in Figure 5.11 (4 passes) and Figure 5.12 (20 passes) [140]. The images are in focus at the field mirror, which also has cut-offs for the incoming and leaving light. The mirrors are mounted in metallic boxes attached to the recipient. The combination of the relatively thick metallic walls of the vessel and their high thermal conductivity provides minimal thermal distortion of the multi pass cell due to changes of the mirror spacing. In a well-designed White cell all the light travels the same known distance through the gaseous medium within 1 % for beams up to 16 degrees wide.

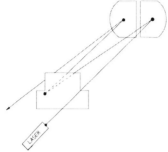

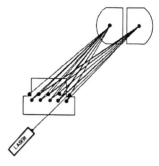

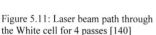

Figure 5.11: Laser beam path through
the White cell for 4 passes [140]

Figure 5.12: Laser beam path through
the White cell for 20 passes [140]

After leaving the plasma reactor the infrared diode laser beam first passed through a grating
monochromator that serves as a mode filter. In addition, the disturbing infrared emission of
the discharge was blocked that way. Depending on its blaze angle and the wave number
range under investigation the grating of the monochromator was exchanged to improve the
intensity of the transmitted laser beam. The laser signal was recorded using a HgCdTe
infrared detector mounted in a liquid nitrogen dewar. The absorption signal of the detector
was amplified and transferred to a PC system. Wedged beam combiners and windows are
used in the complete optical arrangement to avoid unwanted etalon background structure.

5.4.3   Rapid Scan Software

For the measurements a rapid scan software using direct absorption was used to determine
the absolute concentrations of the molecular species, with a time resolution of 2...0.5 Hz.
For data acquisition an advanced form of sweep integration was applied which is carried out
by a special software package [46,141,142]. The hardware of the tunable diode laser system
consists of a PC combined with a high speed (300 kHz) analogue-to-digital board (Scientific
Solutions Lab Master) and the diode laser controller (Laser Analytics Inc., Model L5831).

The software sweeps the laser frequency over the full infrared transition or group of
transitions, then integrates the area under the transitions using non-linear least squares
fitting to the known spectral line shapes and positions. Absolute species concentrations are
returned from the non-linear least squares fits. For time resolved measurements under the
present conditions the stream mode of data acquisition was applied. Frequency modulation
(FM) or second derivative techniques are not used in this program, because the clear

connection between the direct absorption spectrum and the species concentration is preferred. For non time resolved measurements long average times lead to improved signal to noise ratios which make second derivative measurements unnecessary.

There are several advantages to this sweep integration approach. First, absolute species concentrations are returned from the nonlinear least square fits so that external calibration is not required. The species concentrations may be determined from the absolute spectroscopic data available from the HITRAN [117] or GEISA [119] databases or from user supplied data. Second, the line shape functions are known from theory and can be precisely calculated. Finally, this detailed understanding of the expected line shapes and positions allows the operator to easily monitor complex and overlapping spectral features using "fingerprint fitting" where all the individual spectral lines which contribute to a blended absorption feature are included in the fit. The combination of spacings and strengths can result in a unique signature or fingerprint of the molecular species. Monitoring several transitions for one species can enhance sensitivity and is sometimes unavoidable especially for larger molecules like ethane. In addition, fingerprint fitting allows one to monitor multiple species simultaneously, since overlapping lines can be used, and it even can allow the fitting of unknown lines which overlap the desired spectrum as a method of removing background absorption from unknown species.

The basic functions of the program for controlling the diode laser are as follows: The software creates a waveform, which is used to modulate the frequency of the laser. The digital waveform is converted to an analogue voltage at a rate of 300 kHz using a 12 bit digital to analogue converter. E.g. if 150 points are used to represent the waveform each frequency sweep is 500 µs in duration. The current is varied linearly for a given time period, which causes a variation of the frequency of the laser light that the diode emits. Then the current is lowered to a value, which is below the threshold for laser emission. This provides a measurement of detector output in the absence of laser at the end of the scan.

The detector output voltage is sampled by a second analogue-to-digital converter on the same data acquisition board. Every individual spectra are automatically transferred to the computer's memory. The program divides the data in the memory into individual sweeps and averages the sweeps to produce one resultant spectrum. The resultant spectra were analysed spectroscopically to determine the concentrations that absorbs in this special spectral window. The column densities were displayed on screen, saved to disk and analysed.

5.5   Investigations of Hydrogen-Nitrogen Plasmas Containing Methane or Methanol

5.5.1   Introduction

A representative TDL absorption spectrum of a microwave plasma (H_2-Ar-N_2-CH_3OH, p= 1.5 mbar), using the multiple pass arrangement, with lines due to methanol and the methyl radical produced in the plasma is shown in Figure 5.13. This spectrum was recorded using the multiple pass arrangement with 20 passes through the discharge chamber. The dotted lines of N_2O come from a calibration gas cell placed in the infrared light path.

The identification of lines and the measurement of their absolute positions were carried out using well documented reference gas spectra and an etalon of known free spectral range for interpolation [115,117,120] as described in chapter 3.4. Table 5.1 and Table 5.2 give the spectral positions and line strengths of the absorption lines shown in Figure 5.13 and used for concentration measurements. The investigated species listed in Table 5.2 were chosen since they were expected to be the dominant ones in the plasma chemistry and their line positions and line strength were known.

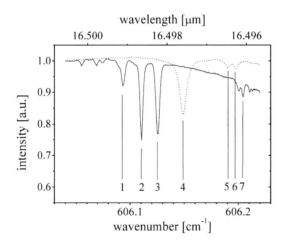

Figure 5.13: TDL absorption spectra of some methyl and methanol lines in a H_2-Ar-N_2-CH_3OH microwave discharge (1,3,7 – CH_3; 2 – CH_3OH; 4,5,6 – N_2O). The dotted lines due to N_2O are from a reference gas cell placed in the beam path.

Line number	Species	Line	Line position [cm^{-1}]	Line strength [cm / molecule]	Ref.
1	CH$_3$	Q(2,2)	606.09172		[41]
2	CH$_3$OH		606.110		
3	CH$_3$	Q(3,3)	606.12032	5.8×10^{-19}	[2,41,143]
4	N$_2$O	R(19)	606.146055	6.05×10^{-22}	[117]
5	N$_2$O	R(19)	606.188708	5.6×10^{-23}	[117]
6	N$_2$O	R(19)	606.197530	5.6×10^{-23}	[117]
7	CH$_3$	Q(1,1)	606.20164		[41]

Table 5.1: Properties of identified lines shown in Figure 5.13

Species	Line	Line position [cm^{-1}]	Line strength [cm / molecule]	Ref.	Remarks
CH$_3$OH		1347.572	$\approx 1.0 \times 10^{-21}$		Calibrated against known CH$_3$OH concentrations in the reactor
CH$_4$		1347.05429	5.824×10^{-20}	[117]	Used for lower concentrations
		1347.19515	3.145×10^{-22}	[117]	Used for higher concentrations
NH$_3$		948.23206	1.386×10^{-19}	[117]	
HCN	R(23)	785.38940	1.040×10^{-21}	[117]	Used for lower concentrations
	R(24)	785.57666	1.186×10^{-20}	[117]	Used for higher concentrations
CH$_3$	Q(3,3)	606.12032	5.80×10^{-19}	[2,41]	Calculated from Q(8,8)
C$_2$H$_2$	R(23)	785.41940	1.160×10^{-19}	[117]	Used for power variations
	R(7)	1347.1631	1.329×10^{-19}	[117]	Used for gas flow variations
C$_2$H$_4$		947.9829	3.509×10^{-20}	[117]	
		948.0020	2.682×10^{-20}	[117]	
C$_2$H$_6$		2993.361...	1×10^{-22}...	[117]	Effective line strength integrated over ≈ 40 lines, calibrated against reference gas cells
		2993.477	1.98×10^{-20}		
CH$_2$O		1777.3196	1.70×10^{-20}	[2]	
CN	R(6)	2067.9114	2.0×10^{-19}	[144]	see chapter 5.6
	R(9)	2078.3001			
C$_2$N$_2$	P(24)	2149.99048	1.15×10^{-20}	[120]	

Table 5.2: Infrared absorption lines used for TDLAS identification and molecular concentration measurements

5.5.2   Experimental Results

5.5.2.1   General features

The concentrations of species measured at N_2 flows of up to 455 sccm by TDLAS are shown for flowing conditions in Figure 5.14 (for methane) and Figure 5.15 (for methanol) respectively. The corresponding results for static conditions are shown in Figure 5.16 and Figure 5.17. For clarity the data for some of the hydrocarbon products are shown separately in the right hand panels of each figure. (Dotted lines have been included as a guide to the eye.) The data were acquired at constant microwave power of 1.5 kW and total pressure (1.5 mbar). The flow used in these figures for methane and methanol (40 sccm) was selected to represent typical flows used in practical CVD reactors. The initial hydrocarbon precursor concentrations are displayed by a full line in the upper part of the respective figure.

Figures 5.14 to 17 give an overview of the mass balance and degree of dissociation, as well as the product concentrations which range over five orders of magnitude. Some useful generalisations follow from these figures. There are some differences between the species concentration profiles measured in flowing (Figure 5.14 and Figure 5.15) and static (Figure 5.16 and Figure 5.17) conditions. The same products are not always found for both precursor hydrocarbons. Thus C_2N_2 was detected in methane plasmas, but not in plasmas with methanol admixture, where CH_2O was found.

Not surprisingly, the hydrocarbon species appear even in the absence of nitrogen. The most noticeable nitrogen-dependent changes occurred for the C-2 hydrocarbons at low nitrogen concentrations for static conditions. The wide range of concentrations measured for the different species provides a convenient means for detailed discussions below. The highest concentrations are for the precursor molecules and the major products HCN and NH_3 followed by the stable and unstable hydrocarbon products (C_2H_2, C_2H_4, C_2H_6 and CH_3). Both HCN and NH_3 concentrations rise under flowing conditions rapidly as the proportion of N_2 reaches 25 % and then are approximately constant. In contrast to the other species for the admixture of methane C_2N_2 rise steadily with the amount of nitrogen admixture in the plasma up to the maximum proportion of nitrogen. C_2N_2 was the species with the lowest concentration of all those studied.

In the case of methanol admixture methane is the dominant hydrocarbon product (Figure 5.15, Figure 5.17).

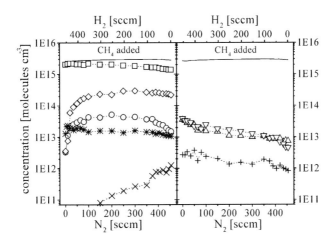

Figure 5.14: Molecular concentrations in a methane containing discharge under flowing conditions over the nitrogen flow rate. For clarity the produced hydrocarbons are presented separately ($\square$ - CH_4, $\diamond$ - HCN, $\bigcirc$ - NH_3, $\ast$ - CH_3, $\times$ - C_2N_2, $\triangle$ - C_2H_2, $+$ - C_2H_4, $\triangledown$ - C_2H_6)

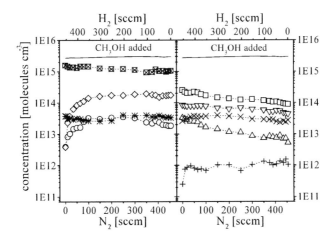

Figure 5.15: Molecular concentrations in a methanol containing discharge under flowing conditions over the nitrogen flow rate. For clarity the produced hydrocarbons are presented separately ($\boxtimes$ - CH_3OH, $\diamond$ - HCN, $\bigcirc$ - NH_3, $\ast$ - CH_3, $\square$ - CH_4, $\times$ - CH_2O, $\triangle$ - C_2H_2, $+$ - C_2H_4, $\triangledown$ - C_2H_6)

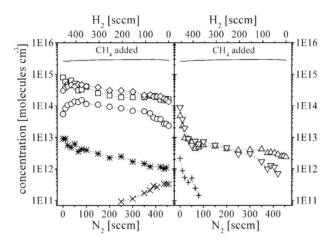

Figure 5.16: Molecular concentrations in a methane containing discharge under static conditions over the nitrogen flow rate. For clarity the produced hydrocarbons are presented separately ($\square$ - CH_4, $\diamond$ - HCN, $\bigcirc$ - NH_3, $\ast$ - CH_3, $\times$ - C_2N_2, $\triangle$ - C_2H_2, $+$ - C_2H_4, $\triangledown$ - C_2H_6)

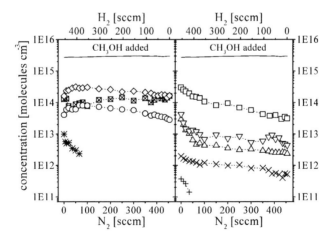

Figure 5.17: Molecular concentrations in a methanol containing discharge under static conditions over the nitrogen flow rate. For clarity the produced hydrocarbons are presented separately ($\boxtimes$ - CH_3OH, $\diamond$ - HCN, $\bigcirc$ - NH_3, $\ast$ - CH_3, $\square$ - CH_4, $\times$ - CH_2O, $\triangle$ - C_2H_2, $+$ - C_2H_4, $\triangledown$ - C_2H_6)

In Figure 5.18 the transition between flowing and static conditions is shown for the first eighty seconds after the reactor had been isolated. This figure demonstrates the temporal development of the molecular concentrations for a pre-isolation flowing gas mixture of 405 sccm H_2 + 60 sccm Ar + 50 sccm N_2 + 40 sccm CH_4 or CH_3OH. The complete temporal changes of molecular concentrations under first flowing conditions (fc) and later static discharge conditions (sc) are presented in the appendix, Figure 10.1 and Figure 10.2.

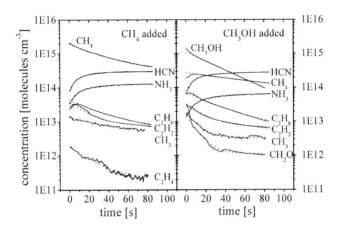

Figure 5.18: Temporal development of molecular concentrations under static conditions in a microwave discharges containing methane or methanol. The initial hydrogen gas flow rate was 405 sccm.

Detecting the species variation in real time opens up a novel approach to obtaining insight into the plasma chemical reactions. The concentration of the precursor molecules decreases continuously with time. From the temporal behaviour of the product species it is clear that HCN and NH_3 are the final stable products. An interesting feature is that before the concentration of all other product hydrocarbons (CH_4, C_2H_2, C_2H_6, CH_2O and CH_3) has decreased with time, the densities of CH_4, C_2H_2 and C_2H_6 show slight maxima after a few seconds. This indicates that these hydrocarbons are intermediate products of the conversion process starting with CH_4 and ending with HCN or NH_3. Recently a comparable phenomenon was observed in a time resolved laser study of the hydrocarbon concentrations in H_2-CH_4 surface wave plasmas for static discharge conditions [12]. Concentration maxima were found for C_2H_2 and C_2H_4, while C_2H_6 was identified as the final product of the plasma chemical conversion.

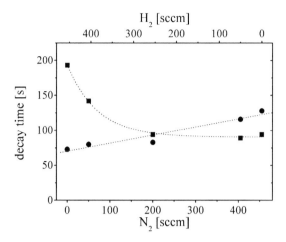

Figure 5.19: The time for an one order of magnitude decay in concentration of methane (■) and methanol (●) as a function of the hydrogen content of the source gas mixture

The slopes of the concentration decay of the precursor hydrocarbons were analysed as well. From the example, presented in Figure 5.18, it is clear, that the temporal decay of the concentration of the precursor methanol is more pronounced than for methane. For a wider range of source gas mixtures the time for a decay in concentration of one order of magnitude is compared for methane and methanol in Figure 5.19. It can be seen that a higher amount of hydrogen leads to a more effective fragmentation of methanol while methane shows an opposite behaviour.

5.5.2.2   Degree of Hydrocarbon Precursor Dissociation

Knowledge of the degree of dissociation of the hydrocarbon precursors is necessary in order to study the conversion of the added gas mixture into products and for understanding the mass balance of the plasma processes. Figure 5.20 shows the degree of dissociation of methane and methanol at increasing nitrogen flow rates under flowing (open symbols) and static (closed symbols) conditions for 7.2 % hydrocarbon content in the flow (40 sccm). It can be seen that increasing nitrogen flow rates have a more noticeable effect for methanol than for methane plasmas. For flowing conditions, with small flows of nitrogen, the degree of dissociation of methanol was found to be about twice as high as for methane. The values for both precursor molecules increase nearly linearly with nitrogen admixture. Under static conditions the degree of dissociation is higher, as expected, and reaches 98 % for methanol and 93 % for methane plasmas. At other precursor hydrocarbon admixtures (namely mixtures containing 0.9, 2.7 or 4.5 % of methane or methanol) the degree of dissociation showed, within error limits, a similar dependence on the partial pressure of nitrogen (see Figure 10.3 in the appendix). A similar difference between the degree of dissociation of methane and methanol was also reported in ref. 2, where it was suggested that different cross sections for electron impact dissociation and a changed electron energy distribution function were responsible.

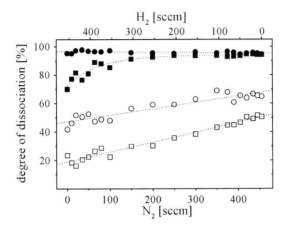

Figure 5.20: Degree of dissociation of methane (□,■) and methanol (○,●) under flowing (open symbols) and static (solid symbols) discharge conditions for various H_2 / N_2-ratios. The flow rates of methane and methanol were 40 sccm.

5.5.2.3   Mass Balance Considerations

The carbon mass balance in methane or methanol containing discharges under static conditions versus the nitrogen flow rate is shown in Figure 5.21 and Figure 5.22, respectively. The upper line represents the sum of all carbon containing molecules. It is easy to see that in the methane case between 10 and 20 % of the available carbon appears as HCN. Only at low amounts of nitrogen do C_2H_2 and C_2H_6 contribute measurably to the mass balance. In the methanol case the methane produced contributes additionally of the order of a few per cent (Figure 5.22). Since generally these plasmas tend to produce layers, it is reasonable to suppose that depositions on the reactor walls but also the production of higher hydrocarbons or other non-measured carbon containing stable or radical species act also as sinks for carbon.

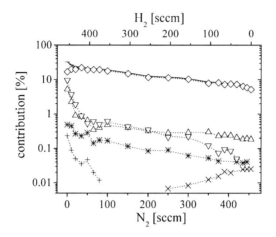

Figure 5.21: Carbon mass balance in a typical methane containing discharge under static conditions versus the nitrogen flow rate. The line at the top of the figure represents the sum of all carbon containing molecules (contribution [%] of $\diamond$ - HCN, $\triangle$ - C_2H_2, $+$ - C_2H_4, $\triangledown$ - C_2H_6, $\ast$ - CH_3, $\times$ - C_2N_2)

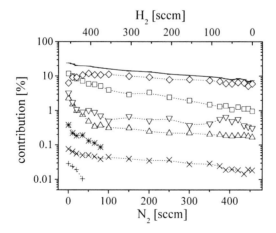

Figure 5.22 Carbon mass balance in a typical methanol containing discharge under static conditions over the nitrogen flow rate. The line at the top of the figure represents the sum of all carbon containing molecules (contribution [%] of $\diamond$ - HCN, $\square$ - CH_4, $\triangle$ - C_2H_2, $+$ - C_2H_4, $\triangledown$ - C_2H_6, $\ast$ - CH_3, $\times$ - CH_2O)

5.5.2.4   Main Products of the Discharge (HCN and NH₃)

The concentrations of ammonia and hydrogen cyanide are given as a function of the nitrogen flow rate and for 7.2 % hydrocarbon admixture in Figure 5.23 and Figure 5.24. Generally the NH_3 and HCN concentrations increased up to about 100 sccm nitrogen admixture, showed a flat maximum and decreased at nitrogen admixtures higher than 400 sscm, caused by the reduced availability of hydrogen (below 20 %). The highest concentration of ammonia of about 1.5×10^{14} molecules cm^{-3} was found in the methane case under static discharge conditions. For 100 sccm nitrogen admixture the dependencies of the NH_3 and HCN concentrations on the precursor content in the source gas are shown in Figure 5.25 and Figure 5.26, respectively. Both species showed a tendency to higher concentrations with an increased admixture of the hydrocarbon precursors. For HCN this effect is more pronounced. The highest concentration of HCN of about 5×10^{14} molecules cm^{-3} was measured in the methane case under static conditions, representing about 20 % of the carbon available in the discharge. For the admixture of methanol lower HCN concentrations were measured. This can be caused by the higher number of carbon containing species in discharges with methanol. Figure 10.6 and Figure 10.7 in the appendix give an overview of the HCN concentration behaviour for all four hydrocarbon percentages investigated for flowing (Figure 10.6) as well as for static discharge conditions (Figure 10.7).

Another interesting effect is the small dependency of the NH_3 concentration on the precursor admixture. The main part of NH_3 is obviously formed by hydrogen and nitrogen in the gas flow with small contribution of the hydrocarbon precursor molecule only. This was proofed by a measurement without any hydrocarbon admixture in the gas flow. These results are given in Figure 5.25 as well as in Figure 10.4 and Figure 10.5, where the change of the NH_3 concentration for various hydrocarbon flows of 0...7.2 % under flowing and static conditions is shown. An interesting effect is that with methane in the gas phase a significant increase of the NH_3, in particular under static discharge conditions, was measured in comparison to the case without any hydrocarbon added. Obviously a high concentration of methane or a related species like the methyl radical improves the formation of NH_3 in the discharge.

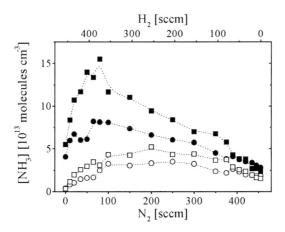

Figure 5.23: Variation of the NH$_3$ concentration with the nitrogen flow rate in hydrocarbon containing discharges under flowing conditions (open symbols) and static conditions (closed symbols). ($\square$,$\blacksquare$ - 40 sccm CH$_4$, $\bigcirc$,$\bullet$ - 40 sccm CH$_3$OH)

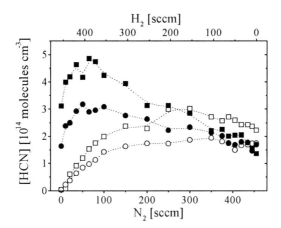

Figure 5.24: Variation of the HCN concentration with the nitrogen flow rate in hydrocarbon containing discharges under flowing conditions (open symbols) and static conditions (closed symbols). ($\square$,$\blacksquare$ - 40 sccm CH$_4$, $\bigcirc$,$\bullet$ - 40 sccm CH$_3$OH)

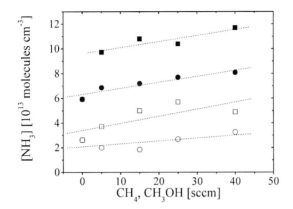

Figure 5.25: Variation of the NH₃ concentration measured by TDLAS as a function of the methane (□,■) or methanol (○,●) flow rate added to H₂-Ar-N₂ plasmas under flowing (□,○) or static (■,●) discharge conditions (N₂ – 100 sccm)

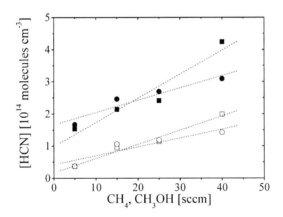

Figure 5.26: Variation of the HCN concentration measured by TDLAS as a function of the methane (□,■) or methanol flow rate (○,●) added to H₂-Ar-N₂ plasmas under flowing (□,○) and static (■,●) discharge conditions (N₂ – 100 sccm)

5.5.2.5   The Methyl radical (CH$_3$)

The methyl radical is one of the first species formed in the decomposition process of hydrocarbons precursors. Therefore, it is of importance for an improved knowledge of the chemical reaction kinetics. Hence, the hypothesis that the methyl free radical is the one of the most likely precursors for forming thin carbon films has gained acceptance. The methyl radical has therefore been widely studied in carbon film deposition plasmas and is one of the main species of interest in the present investigation too.

The electronic spectrum of the methyl radical was first detected in the vacuum ultraviolet by Herzberg and Shoosmith in 1956 [145], and later analysed by Herzberg [146]. Over a decade later Tan, Winer and Pimentel produced CH$_3$ by flash photolysis of CH$_3$I and recorded its low resolution infrared spectrum using a dispersion spectrometer [225]. In these pioneering IR experiments the sensitivity was, not surprisingly, rather low. Much higher sensitivity and resolution was achieved later by Hirota and co-workers using infrared lasers, in particular applying TDLAS [41,42,147].

While the methyl radical is still the subject of basic research in recent years it has been widely studied in applied investigations [148-151]. Tachibana and co-workers measured the rate of dissociation of CH$_4$ in an RF methane plasma by infrared laser absorption spectroscopy and predicted that CH$_3$ should be the most abundant radical in the plasma [152]. Later Celii and co-workers detected the absorption spectrum of the methyl radical in a hot filament CVD reactor using TDLAS and made rough estimates of the concentrations of CH$_3$, C$_2$H$_2$, CH$_4$, and C$_2$H$_4$ near the filament [153].

Measurements for detecting the methyl radical in absorption are usually performed either at 216 nm in the ultraviolet or in the infrared region near 606 cm^{-1}. The infrared TDLAS technique has proven to be more useful since it is also possible to measure concentrations of related species provided they are active in the mid infra red. Wormhoudt demonstrated this flexibility by measuring CH$_3$ and C$_2$H$_2$ in absorption in a CH$_4$ / H$_2$ RF discharge using a long path cell [154]. An important study for quantifying the concentrations of the methyl radical was the determination of the line strength for the Q(8,8) line of CH$_3$ at 608.3 cm^{-1} by Wormhoudt and McCurdy [143]. Goto and co-workers have published numerous studies of methyl and methanol concentrations in RF and ECR plasmas under different conditions e.g. investigating the influence of rare gases on the plasma. They have also combined infrared absorption with emission spectroscopy, and investigated the effect of water vapour on the methyl radical concentration in argon / methane and argon / methanol RF plasmas using TDLAS [4,155-159].

Recent improvements like more powerful diode lasers, the application of multiple pass optics to a planar microwave plasma source or the use of computer based sweep integration and averaging techniques led to a significant improvement of sensitivity and to improved signal to noise ratios like presented in Figure 5.13.

Depending on the discharge conditions the measured concentrations of the methyl radical ranged from 1×10^{12} to 6×10^{13} molecules cm^{-3}, as can be seen from Figure 5.14-17. When nitrogen was added to plasmas containing methane or methanol as precursor, the methyl radical concentration was found to decay slightly as the flow rate of N_2 increased. Figure 5.27 and 5.28 show the results for a wide range of fixed hydrocarbon precursor flows. The methyl radical concentration can be seen to be slightly higher for methanol than for methane (Figure 5.27) in particular at 40 sccm (Figure 5.28), but also, though less pronounced, at lower precursor flow rates. Figure 5.29 shows the variation of the methyl radical concentration as a function of added methane or methanol and in the absence of nitrogen. Flowing conditions give much higher concentrations than static conditions. A similar behaviour was reported in ref. 2. The relative dependence of [CH$_3$] on the added precursor flow, and even the ratio of the methyl concentrations of about 4 between fc and sc in the case of methanol for 40 sccm, is quantitatively reproduced compared with the earlier study [2] within the error limits. However, it should be noted that the absolute values of the methyl concentration in the absence of nitrogen given here are more than one order of magnitude higher than reported for the same conditions earlier [2]. The reason for this discrepancy is possible due to the different methods of measurement namely the use of diode laser current modulation and measuring second derivative line shapes (calibrated against N_2O absorption lines [2]) in the earlier study. However, concentrations are known to vary in these reactors even for the same starting conditions. As already described above and shown in Figure 5.13, the present study uses direct absorption spectroscopy, including for methyl radical concentration measurement. This became possible since

(i)   laser diodes with higher intensity are now available leading to improved signal to noise ratios,

(ii)   sweep integration has been applied and,

(iii)   the optical path length was greatly increased.

Recently the infrared TDLAS method for detection of the methyl radical has been validated by comparison with uv absorption measurements at 216 nm in the same reactor, leading, within error limits, to the same results [160].

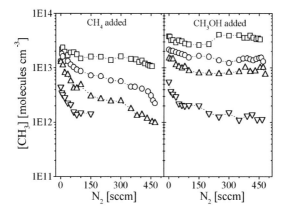

Figure 5.27: Variation of the CH₃ concentration with the nitrogen flow rate under flowing discharge conditions for hydrocarbon precursor admixtures of 40 sccm (□), 25 sccm (○), 15 sccm (△) and 5 sccm (▽) and various H_2 / N_2-ratios

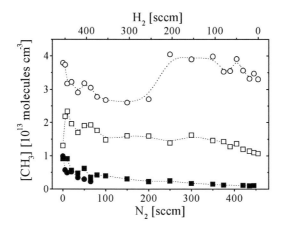

Figure 5.28: Variation of the CH₃ concentration with the nitrogen flow rate in hydrocarbon containing discharges under flowing conditions (open symbols) and static conditions (closed symbols). (□,■ - 40 sccm CH₄, ○,● - 40 sccm CH₃OH) and various H_2 / N_2-ratios

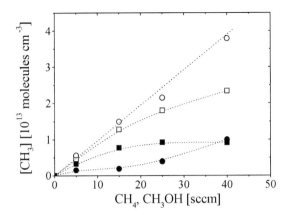

Figure 5.29: Variation of the CH_3 concentration as a function of methane ($\square$,$\blacksquare$) or methanol ($\bigcirc$,$\bullet$) added to H_2-Ar plasmas under flowing (open symbols) and static (solid symbols) discharge conditions. The data have been fitted to a second-order polynomial.

5.5.2.6 Other Carbon containing products

Concentration behaviours of other carbon containing species are presented in Figure 5.30 for flowing conditions and in Figure 5.31 for static conditions. Some differences between the two hydrocarbon precursors are obvious:

(I)   For methanol admixtures, methane is the dominant hydrocarbon product (see Figure 5.15 und 5.17).

(II)   While for methane admixtures the concentrations of C_2H_2, CH_3 and C_2H_6 are comparable for all H_2 / N_2-ratios, in the methanol case for higher nitrogen contents several times more of CH_3 and C_2H_6 is formed.

(III) In addition, H_2CO was observed for methanol admixture with concentrations in the range of the product hydrocarbons.

(IV) C_2N_2 is formed as another carbon containing molecule with methane in gas flow and rather low concentration values.

Obviously, the dissociation of methanol and the formation of various hydrocarbons are more efficient for methanol as hydrocarbon precursor. A wider number of different species is formed leading to a more complicate chemical reaction system. In the appendix in Figure 10.10 – Figure 10.19 the concentrations of the hydrocarbon products are shown for various precursor percentages in the premixed gas flow.

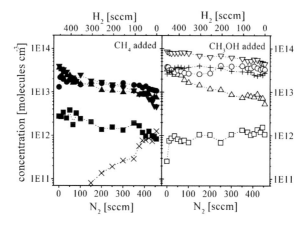

Figure 5.30: Molecular concentrations of hydrocarbons produced under flowing discharge conditions for methane (closed symbols) and methanol (open symbols) admixture over the nitrogen flow. ($\blacktriangle,\triangle$ - C_2H_2, $\blacksquare,\square$ - C_2H_4, $\blacktriangledown,\triangledown$ - C_2H_6, $\bullet,\circ$ - CH_3, $\times$ - C_2N_2, $+$ - CH_2O)

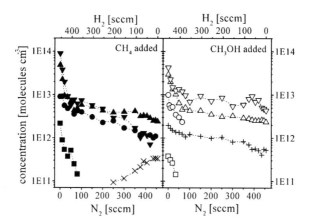

Figure 5.31: Molecular concentrations of hydrocarbons produced under static discharge conditions for methane (closed symbols) and methanol (open symbols) admixture over the nitrogen flow. ($\blacktriangle,\triangle$ - C_2H_2, $\blacksquare,\square$ - C_2H_4, $\blacktriangledown,\triangledown$ - C_2H_6, $\bullet,\circ$ - CH_3, $\times$ - C_2N_2, $+$ - CH_2O)

5.5.2.7   Fragmentation and Conversion rates

To gain further insight into plasma chemical conversion the experimental data concerning methane and methanol dissociation was used to estimate absolute fragmentation rates of CH_4 and CH_3OH and conversion rates to the measured main products CH_4, HCN and NH_3. These rates are normalised on the discharge power.

The fragmentation rate R_F of the hydrocarbon precursor molecules methane and methanol is introduced by analogy to [11] as

$$R_F = \Phi_P \ 1/60 \ D/100 \ N_o/P \qquad (5.10)$$

where R_F has the units of molecules J^{-1}, Φ_P is the precursor flow (sccm), D is the degree of dissociation of the precursor molecules given in percent, N_o is the number of molecules per cm^{-3} at norm conditions (2.69 x 10^{19} molecules) and P is the power (W).

The conversion rate R_C to plasma product molecules is expressed analogously as

$$R_C = n_{molecule} \ \Phi_{total}/60 \ 10^3/p \ 1/P \qquad (5.11)$$

where R_C has the units of molecules J^{-1}, $n_{molecule}$ is the measured molecular concentration in molecules cm^{-3}, Φ_{total} is the total gas flow in sccm, p is the pressure in mbar and P is the power (W).

Figure 5.32 shows the respective rate of fragmentation of methane and methanol for increasing nitrogen flows containing 7.2 % of the precursor gas and for flowing conditions. In nitrogen-free plasmas the rate of fragmentation of methanol ($R_F(CH_3OH) \approx 6$ x 10^{15} molecules J^{-1}) is twice as high as for methane ($R_F(CH_4) \approx 3$ x 10^{15} molecules J^{-1}). On adding nitrogen to the plasma, while keeping the total gas flow and the discharge pressure constant, i.e., varying the ratio of hydrogen to nitrogen, both rates of fragmentation increase nearly linearly up to about 7 x 10^{15} molecules J^{-1} for methane and about 9 x 10^{15} molecules J^{-1} for methanol. Similar results have been obtained for other admixtures of hydrocarbon precursors. In general smaller admixtures of hydrocarbons resulted in lower fragmentation rates. The values of the fragmentation rates of methane at high nitrogen flow rates presented here are near to those found in a H_2-CH_4 surface wave discharge at a comparable high total gas flow rate [11].

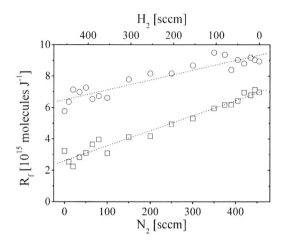

Figure 5.32: The fragmentation rate, R_F, of methane ($\square$) (fc) and methanol ($\bigcirc$) (fc) as the flow rate of nitrogen was increased (CH₃OH, CH₄: 40 sccm)

Figure 5.33 shows the rates of conversion into HCN and NH₃ in H₂-Ar-N₂ microwave plasmas containing methane at flowing conditions as well as the corresponding conversion rates, including that to methane, for plasmas containing methanol. The rates of conversion to particular reaction products are based on the measurements of their absolute concentrations. They range between 10^{14} to 10^{15} molecules J⁻¹. The conversion rate to HCN is about one order of magnitude higher than that of NH₃. H₂-Ar-N₂ microwave plasmas with methane as precursor tend to show slightly higher conversion rates compared to the methanol case. Both rates of conversion rise steeply to a maximum at the ratio of hydrogen to nitrogen of 1:1. Thus, methane was detected in methanol plasmas, its conversion rate decreases from $12 - 4 \times 10^{14}$ molecules J⁻¹ with increasing nitrogen flow rate.

Conversion rates to hydrocarbons are presented in Figure 5.34. While the values for C₂H₂ and C₂H₄ are similar for methane and methanol as precursor, the conversion rate to C₂H₆ is in the methanol admixture case by a factor of about 4 higher. A major difference in comparison with [11] is that in the present work for high flow rates and methane as precursor hydrocarbon similar conversion rates to C₂H₂, C₂H₄ and C₂H₆ were determined. These values are here by a factor of about five smaller. This is mainly caused by the additional conversion to HCN and NH₃ under the presence of nitrogen in the actual gas mixtures.

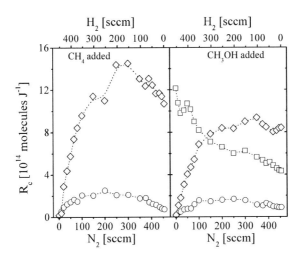

Figure 5.33: The conversion rate, R_C, to HCN ($\diamond$), to NH$_3$ ($\bigcirc$) and to methane ($\square$) under flowing discharge conditions for various H$_2$ / N$_2$-ratios (CH$_3$OH, CH$_4$: 40 sccm)

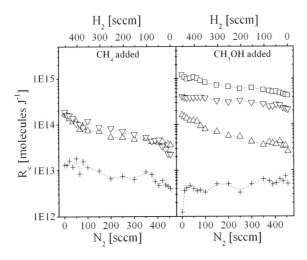

Figure 5.34: The conversion rate, R_C, to CH$_4$ ($\square$), to C$_2$H$_2$ ($\triangle$), to C$_2$H$_4$ ($+$) and to C$_2$H$_6$ ($\triangledown$) under flowing discharge conditions for various H$_2$ / N$_2$-ratios (CH$_3$OH, CH$_4$: 40 sccm)

5.5.3 Plasma Chemistry Modelling for Hydrogen-Nitrogen-Methane Discharges

5.5.3.1 The Chemical Reaction Kinetics Model

Modelling of the chemistry of species rich molecular plasmas containing hydrocarbon has been reported extensively [e.g. 2,56,161-167]. In the majority of examples, methane is the precursor. In the present work, plasmas containing methane or methanol as hydrocarbon precursor molecules are in the centre of interest. The separation of a hydrogen atom from methane, or of the OH group from methanol, by electron impact are processes with large rate coefficient [167] but precise data for typical conditions in the microwave plasma under investigation are scarce. It was found in the present study, and by others workers [2], that methanol is dissociated to a greater extent than methane, but modelling of methanol containing plasmas has not so far been reported. Nevertheless, the C-O bond in the methanol molecule is weaker than the C-H bond in methane, and the methyl radical is one of the first products of electron impact dissociation in both cases. Hence, chemical reaction schemes involving methane and methanol as precursor molecules might be expected to have some common features.

In order to achieve as comprehensive and precise a model of a particular discharge it should ideally include three components namely a plasma model, a plasma chemistry model and a surface model. Each of these parts can be described by equations which illustrate characteristic temporal and spatial variations. Plasma models mainly describe the properties of the charged particles using transport and kinetic equations as well as Maxwell's equations. Plasma chemistry models deal with the chemical reactions of neutrals using transport and kinetic equations, and surface models take for example the deposition or etching processes on walls into account. In principle, none of these parts are independent of the other, but reasonable simplifications are often necessary due to the limited knowledge of the discharge species properties [221,222].

The aim of the plasma chemical modelling described here is the simulation of trends in the molecular species concentrations for comparison with experimental results in order to obtain information about the main reaction paths. Based on such model calculation the concentrations of molecular species hitherto undetected can be estimated. The chemically active plasmas described here are characterized by a large number of different species present in the discharge volume. In order to achieve a reliable chemical model starting with a reduced number of unknown or uncertain parameters several major simplifications have to be made as follows:

(I)   The distances from the discharge observation zone to the metallic walls of the discharge chamber is large in comparison to the mean free path of the molecules. While the plasma-to-wall distance was a minimum of some millimetres, the mean free path length for molecule-molecule and electron-molecule interactions was estimated to be less than a millimetre under our conditions [56]. Therefore, no wall interactions, in particular deposition, etching or recombination effects, were taken into account.

(II)   Electron impact was assumed to be the major process contributing to the dissociation of molecular species in the discharge. Thermal dissociation reactions, in particular of precursor molecules like of hydrogen and nitrogen, were expected to be negligible.

(III)   This type of microwave discharges is characterized by its low degree of ionisation ($\sim 10^{-5}$). Hence the number of ionised species in the discharge chamber is quite low and for this reason reactions involving ions were not used in the model.

(IV)   The model describes static discharge conditions only. Therefore mass transport and diffusion were not included. The discharge chamber was treated as a single cell with uniform plasma properties.

(V)   In the discharges of interest here the concentration of ions and electrons is in the order of 10^{11} cm^{-3} and the mean electron energy is estimated to be about $1 - 2$ eV (see chapter 5.3.2). Hence the influence of excited species on the reaction chemistry was ignored. The only exception was the low-lying first excited state of CH$_2$, labelled as CH2(1) in comparison to CH2(3) to the triplet ground state. It is usually included in the neutral reaction chemistry of hydrocarbon systems (see e.g. ref. 167,174).

(VI)   Based on the typical low degree of ionisation and on recent probe measurements [64] an electron density of 1×10^{11} cm^{-3} was assumed in the model for calculating the electron impact dissociation reaction rates. A common gas temperature of 1000 K was employed for all plasma chemical reactions (see chapter 5.4.1).

The FACSIMILE program code was used for modelling the plasma chemistry. FACSIMILE is a computer software package for solving chemical kinetic problems in the fields of physical and chemical science and engineering. It was used to solve the system of ordinary differential equations determining the temporal evolution of the 22 species concentrations.

Results of model calculations for a microwave discharge containing hydrogen, argon and methane under conditions similar to the present work were reported in refs. 2 and 56. Model calculations of hydrogen, oxygen, argon and methane mixtures have also been presented earlier. The model described in the present chapter deals with gas mixtures of hydrogen, methane and argon, and with nitrogen instead of oxygen.

Rate coefficients for the electron impact dissociation reactions (see Table 5.3) were either calculated for the specific discharge conditions or estimated if cross sections for these processes were not known. This applies in particular the nitrogen containing products like ammonia or hydro cyanic acid. The accuracy of the model is limited by this shortfall. In addition, the limited knowledge of plasma parameters like the electron density and the electron energy distribution function, in turn limits the accuracy of calculated electron impact dissociation rate coefficients.

The change of total electron impact dissociation rate coefficients versus the reduced field strength amplitude E_0/N is shown in Figure 5.35 [168]. The rate coefficients were determined solving the time-dependent Boltzmann equation for a given electric field strength

$$E(t) = E_0 \cos \omega t \tag{5.12}$$

with respect to the microwave frequency of 2.45 GHz, a total gas density N of 3.998×10^{16} molecules cm^{-3}, a $H_2 / N_2 / Ar / CH_4$ gas mixture of 73.0 / 9.0 / 10.8 / 7.2 % and a gas temperature of 1000 K [168]. E_0 in equation (5.12) is the amplitude of the field. In addition to the selected field parameters and the gas composition atomic data, like cross sections, molecule masses and energy losses due to inelastic collisions were used as input data. Cross sections for electron-molecule (H_2, N_2, CH_y and C_2H_y) and electron-argon collisions were taken from the established literature [169-172]. The electron kinetic equation has been solved up to the periodic state for given electric field using the multiterm method, described in detail in ref. 223. The rate coefficients are averaged over the period of the microwave and given as a function of E_0/N, as shown in Figure 5.35, corresponding to a typical gas mixture as mentioned above.

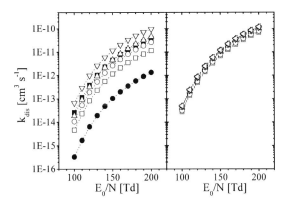

Figure 5.35: Calculated total electron impact dissociation rate coefficients, k_{dis}, for hydrogen, nitrogen and various hydrocarbons as a function of the reduced electric field strength, E_0/N (left panel: ■ - H_2, ● - N_2, □ - CH, ○ - CH_2, △ - CH_3, ▽ - CH_4; right panel: □ - C_2H, ○ - C_2H_2, △ - C_2H_3, ▽ - C_2H_4, ◇ - C_2H_5, ◁ - C_2H_6) [168]

The rate coefficients increase significantly with the reduced field strength. While the values for C_2H_y hydrocarbons (right hand panel in Figure 5.35) were found to be almost independent of the species type, total rate coefficients of lower CH_y hydrocarbons (left hand panel in Figure 5.35) show significant differences. The smallest calculated electron impact dissociation rate coefficients were those for nitrogen.

Based on an estimate of the change of the mean electron energy with the reduced field strength and from experimental investigations of the electron component [64] a reduced field strength E_0/N of about 180 Td was deduced (Table 5.3). The rate coefficients of the electron-molecule molecule collisions of Nitrogen-containing compounds were assumed to be of a comparable order of magnitude in comparison with the values calculated for the hydrocarbons. The data of neutral chemical reaction rate constants were taken from a review of the established literature [173-189]. The accuracy of data for reactions leading to ammonia or hydrogen cyanide is much more limited in comparison with the data for reactions of hydrocarbons, where several reviews are available (see e.g. ref 173,174,177,185,188). For the theoretical analysis of the plasma chemistry, 22 species and 145 reactions were included in the complex reaction kinetic model. The reaction scheme consists of 30 electron impact dissociation reactions and 78 chemical reactions without nitrogen and 37 with. Since the accuracy of known rate coefficients is rather limited, in some cases the rate coefficients had to be adapted to the given discharge conditions. In particular this was necessary for reactions of nitrogen containing compounds and for several hydrogen reactions.

The thermal reactions of neutrals relevant for the model, their rate coefficients at 1000 K and 1.5 mbar and the respective references are listed in Table 5.4. Bimolecular Reaction Rate coefficients are noted here only. Three body recombination reactions involving M are given here with recalculated rate coefficients using the total gas density as value of M.

No.	Reaction	Rate Coefficient [cm³ s⁻¹]	Comment
1	$H2 + E = H + H + E$	2.0E-11	$K(E_0/N)$
2	$N2 + E = N + N + E$	5.9E-13	$K(E_0/N)$
3	$NH + E = N + H + E$	2.0E-12	Estimated
4	$NH2 + E = NH + H + E$	5.0E-12	Estimated
5	$NH2 + E = N + H + H + E$	5.0E-12	Estimated
6	$NH3 + E = NH2 + H + E$	7.0E-12	Estimated
7	$NH3 + E = NH + H + H + E$	7.0E-12	Estimated
8	$CN + E = C + N + E$	2.0E-12	Estimated
9	$C2N2 + E = CN + CN + E$	5.0E-12	Estimated
10	$HCN + E = CN + H + E$	7.0E-12	Estimated
11	$HCN + E = CH + N + E$	7.0E-12	Estimated
12	$CH + E = C + H + E$	6.6E-12	$K(E_0/N)$
13	$CH2(1) + E = CH + H + E$	6.8E-12	$K(E_0/N)$
14	$CH2(1) + E = C + H + H + E$	6.0E-12	$K(E_0/N)$
15	$CH2(3) + E = CH + H + E$	6.8E-12	$K(E_0/N)$
16	$CH2(3) + E = C + H + H + E$	6.0E-12	$K(E_0/N)$
17	$CH3 + E = CH2(3) + H + E$	1.2E-11	$K(E_0/N)$
18	$CH3 + E = CH + H + H + E$	1.0E-11	$K(E_0/N)$
19	$CH4 + E = CH3 + H + E$	2.8E-11	$K(E_0/N)$
20	$CH4 + E = CH2(3) + H + H + E$	1.4E-11	$K(E_0/N)$
21	$C2H + E = C + C + H + E$	3.4E-11	$K(E_0/N)$
22	$C2H2 + E = C2H + H + E$	3.8E-11	$K(E_0/N)$
23	$C2H3 + E = C2H2 + H + E$	2.6E-11	$K(E_0/N)$
24	$C2H3 + E = C2H + H + H + E$	2.3E-11	$K(E_0/N)$
25	$C2H4 + E = C2H3 + H + E$	2.9E-11	$K(E_0/N)$
26	$C2H4 + E = C2H2 + H + H + E$	2.5E-11	$K(E_0/N)$
27	$C2H5 + E = C2H4 + H + E$	3.1E-11	$K(E_0/N)$
28	$C2H5 + E = C2H3 + H + H + E$	2.7E-11	$K(E_0/N)$
29	$C2H6 + E = C2H5 + H + E$	3.1E-11	$K(E_0/N)$
30	$C2H6 + E = C2H4 + H + H + E$	2.7E-11	$K(E_0/N)$

Table 5.3: Molecule - electron collision dissociation reactions and their respective rate coefficients

No.	Reaction	Rate Coefficient [cm^3 s^{-1}]	Reference
1	C + N = CN	4.0E-16	Estimated
2	C + N2 = CN + N	1.0E-12	Estimated
3	CH + N = CN + H	2.1E-11	[176],[183],[187]
4	CN + CN = C2N2	1.4E-10	[185]
5	CN + HCN = C2N2 + H	3.3E-11	[183]
6	CH2(3) + N = HCN + H	8.3E-11	[176],[183],[187]
7	CH3 + N = HCN + H2	1.6E-11	[182]
8	C2H3 + N = HCN + CH2(3)	3.3E-11	[183]
9	CH + N2 = HCN + N	5.3E-16	[182],[183]
10	CN + H2 = HCN + H	1.0E-14	Estimated
11	CN + CH4 = HCN + CH3	1.0E-14	Estimated
12	CN + H = HCN	1.0E-14	Estimated
13	CN + CN = HCN + HCN	1.0E-14	Estimated
14	NH + CH = HCN + H	8.3E-11	[187]
15	NH + CH2(3) = HCN + H + H	5.0E-11	[187]
16	NH2 + CH = HCN + H + H	5.0E-11	[187]
17	NH3+ CN = HCN + NH2	1.7E-11	[185]
18	CN + N = N2 + C	3.0E-14	Estimated
19	C2N2 + M = CN + CN + M	4.0E-16	Estimated
20	HCN + H = H2 + CN	5.0E-14	Estimated
21	HCN + CH = C2H2 + N	1.7E-10	[189]
22	HCN + NH2 = CN + NH3	5.0E-11	Estimated
23	N + H2 = NH + H	4.5E-11	[180]
24	NH + H2 = NH2 + H	6.8E-14	Estimated
25	NH + NH = N + NH2	3.7E-12	[181]
26	NH2 + H2 = NH3 + H	1.0E-12	Estimated
27	NH2 + H = NH3	2.0E-11	Estimated
28	NH2 + NH = NH3 + N	8.0E-11	Estimated
29	NH2 + NH2 = NH3 + NH	1.0E-12	Estimated
30	NH + N = H + N2	2.5E-11	[179],[180],[182]
31	NH + H = H2 + N	4.5E-11	[178],[182],[184]
32	NH + NH = H2 + N2	1.7E-13	[178]
33	NH + NH = N2 + H + H	6.6E-11	[184]
34	NH2 + N = N2 + H + H	1.2E-10	[178],[181],[184]
35	NH2 + N = N2 + H2	1.2E-10	[178]
36	NH2 + H = H2 + NH	1.8E-11	[181],[183]
37	NH3 + H = H2 + NH2	1.0E-13	[182-184]
38	C + H = CH	2.0E-11	Estimated
39	C + H2 = CH + H	2.0E-11	Estimated
40	C + H2 = CH2(3)	2.0E-11	Estimated

No.	Reaction	Rate Coefficient [cm^3 s^{-1}]	Reference
41	CH + H2 = CH2(3) + H	2.0E-11	[182]
42	CH2(1) + M = CH2(3) + M	2.1E-10	[183]
43	CH + H2 = CH3	4.1E-11	[173],[188]
44	CH2(1) + H2 = CH3 + H	1.2E-10	[173],[177],[182]
45	CH2(3) + H2 = CH3 + H	5.0E-13	Estimated
46	CH2(3) + CH2(3) = CH3 + CH	8.0E-11	Estimated
47	CH2(3) + H2 = CH4	8.4E-12	Estimated
48	CH3 + H = CH4	3.0E-12	[186]
49	CH3 + H2 = CH4 + H	5.0E-13	Estimated
50	CH2(3) + C = C2H + H	9.0E-10	Estimated
51	CH + CH = C2H + H	9.0E-10	Estimated
52	C2H + H = C2H2	3.0E-10	[177],[186]
53	C2H + H2 = C2H2 + H	5.0E-10	Estimated
54	CH3 + C = C2H2 + H	8.0E-10	Estimated
55	CH + CH = C2H2	2.0E-10	[176],[187]
56	CH2(3) + CH = C2H2 + H	8.0E-10	Estimated
57	CH2(3) + CH2(3) = C2H2 + H + H	1.2E-10	[173],[176]
58	CH2(3) + CH2(3) = C2H2 + H2	1.3E-11	[173],[176],[187]
59	C2H + CH2(3) = C2H2 + CH	3.0E-11	[176],[177]
60	C2H + CH2(1) = C2H2 + CH	3.0E-11	[177]
61	C2H + CH4 = C2H2 + CH3	2.0E-12	[176],[186]
62	C2H2 + H = C2H3	6.3E-14	[173],[188]
63	C2H3 + H2 = C2H4 + H	1.0E-12	Estimated
64	C2H3 + H = C2H4	5.0E-10	Estimated
65	CH4 + C = C2H4	8.0E-12	Estimated
66	CH4 + CH = C2H4 + H	9.9E-11	[165],[182],[189]
67	CH2(3) + CH2(3) = C2H4	1.7E-12	[176],[187]
68	CH3 + CH2(3) = C2H4 + H	7.0E-11	[176],[177],[188]
69	CH3 + CH2(1) = C2H4 + H	3.0E-11	[177],[187]
70	CH3 + CH3 = C2H4 + H2	1.7E-15	[174]
71	C2H3 + CH4 = C2H4 + CH3	8.0E-11	Estimated
72	C2H3 + C2H3 = C2H4 + C2H2	2.0E-10	Estimated
73	C2H4 + H = C2H5	7.8E-12	[174]
74	CH4 + CH = C2H5	1.0E-10	[167]
75	CH3 + CH3 = C2H5 + H	5.6E-14	[173],[188]
76	CH4 + CH3 = C2H5 + H2	1.7E-16	[176]
77	CH3 + CH3 = C2H6	2.5E-11	[174],[176],[187]
78	C2H5 + H2 = C2H6 + H	4.6E-15	[177],[188]
79	C2H5 + H = C2H6	5.0E-11	[186]
80	C2H5 + CH4 = C2H6 + CH3	5.0E-14	Estimated
81	C2H5 + C2H3 = C2H6 + C2H2	8.0E-13	[176],[177]

No.	Reaction	Rate Coefficient [cm^3 s^{-1}]	Reference
82	C2H5 + C2H4 = C2H6 + C2H3	3.0E-16	[177]
83	C2H5 + C2H5 = C2H6 + C2H4	2.4E-12	[176],[187],[188]
84	CH + H = C + H2	2.5E-10	[183]
85	CH2(3) + H = CH + H2	6.6E-11	[164],[174]
86	CH2(1) + H = CH + H2	5.0E-11	[177],[182]
87	CH3 + H2 = CH2(3) + H + H2	5.0E-12	Estimated
88	CH3 + H = CH2(3) + H2	1.0E-13	[183]
89	CH3 + H = CH2(1) + H2	1.0E-15	[173],[188]
90	CH4 + H = CH3 + H2	4.5E-13	[164],[174],[177]
91	CH4 + CH2(3) = CH3 + CH3	1.0E-14	[33]
92	CH4 + CH2(1) = CH3 + CH3	7.1E-11	[177]
93	C2H + H2 = CH2(3) + H	7.0E-11	[175]
94	C2H2 + H = C2H + H2	6.7E-16	[164],[174]
95	C2H3 + CH = C2H2 + CH2(3)	8.3E-11	[183]
96	C2H3 + CH2(3) = C2H2 + CH3	3.0E-11	[176],[177],[186]
97	C2H3 + CH2(1) = C2H2 + CH3	3.0E-11	[177]
98	C2H3 + CH3 = C2H2 + CH4	6.5E-13	[176],[177],[186]
99	C2H3 + C2H = C2H2 + C2H2	1.6E-12	[177],[186]
100	C2H3 + H = C2H2 + H2	2.0E-11	[173],[176],[188]
101	C2H4 + H = C2H3 + H2	1.5E-12	[164],[174]
102	C2H4 + CH3 = C2H3 + CH4	2.5E-14	[173],[188]
103	C2H5 + H = C2H4 + H2	3.0E-12	[176],[177]
104	C2H5 + H = CH3 + CH3	6.0E-11	[176],[177],[188]
105	C2H5 + CH2(3) = C2H4 + CH3	3.0E-11	[177],[186]
106	C2H5 + CH3 = C2H4 + CH4	1.3E-12	[164]
107	C2H5 + C2H = C2H4 + C2H2	3.0E-12	[176],[177]
108	C2H5 + C2H3 = C2H4 + C2H4	1.0E-10	Estimated
109	C2H6 + C2H5 = C2H6 + C2H4 + H	8.6E-14	[188]
110	C2H6 + H = C2H5 + H2	2.1E-12	[174],[177],[183]
111	C2H6 + CH2(3) = C2H5 + CH3	7.7E-14	[186]
112	C2H6 + CH2(1) = C2H5 + CH3	2.0E-10	[177],[183]
113	C2H6 + CH3 = C2H5 + CH4	1.4E-14	[164],[174],[177]
114	C2H6 + C2H = C2H5 + C2H2	6.0E-12	[176],[177]
115	C2H6 + C2H3 = C2H5 + C2H4	4.0E-14	[177]

Table 5.4: Chemical reactions and their respective rate coefficients at T= 1000 K

5.5.3.2 Results

The model calculations of the chemical kinetics described above were performed for a gas mixture of 73.0 % hydrogen, 9.0 % nitrogen, 10.8 % argon and 7.2 % methane under static conditions assuming plasma properties similar to the ones in the planar microwave discharge. The results of model calculations were taken from the temporal development output of FACSIMILE after a period of 60 sec. Therefore they can be compared with experimental results taken after the same time in the discharge. As starting values for the model calculations measurement results of the main plasma compounds, i.e. of ammonia (NH_3), hydro cyanic acid (HCN), the methyl radical (CH_3), methane (CH_4), acetylene (C_2H_2), ethylene (C_2H_4) and ethane (C_2H_6), under flowing discharge conditions were used. Assumptions needed to be made for the unmeasured species. For molecular hydrogen and nitrogen concentrations according to their content in the gas flow were used. The concentration of atomic hydrogen was found to be two orders of magnitude smaller under flowing conditions [190]. For this reason the degree of dissociation of nitrogen was assumed to be of the same order of magnitude as hydrogen. The resulting concentrations were found to be relatively insensitive to the assumed starting concentrations, which were generally rather low compared to the concentrations of the main plasma compounds.

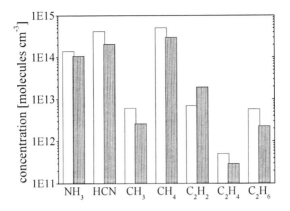

Figure 5.36: Comparison of species concentrations in a representative H_2-N_2-Ar-CH_4-plasma (white – measured by TDLAS, grey – calculated)

The comparison between calculated and measured species concentrations is shown in Figure 5.36. A satisfactory agreement was found for all seven species between measured and calculated concentrations, to within the order of the measurement errors. Discrepancies can be accounted for by the simplifications, which had to be made. Moreover, concentrations of additional species, in particular of important transient species, can be estimated with the aid of the model calculations. The predicted concentrations of several undetected radical species are reported in Table 5.5.

Species:	NH	NH_2	CH	CH_2	C_2H	C_2H_3	C_2H_5
Concentration: [molecules cm^{-3}]	6×10^{12}	3×10^{11}	8×10^{10}	1×10^{12}	8×10^{6}	3×10^{9}	5×10^{10}

Table 5.5: Predicted concentrations from the model for so far undetected species

Using the present absorption experiment with multiple pass arrangement, some of these radicals, in particular NH, NH_2 or CH_2, should be observable. But the lack of line strengths data for these species makes the determination of absolute concentrations impossible.

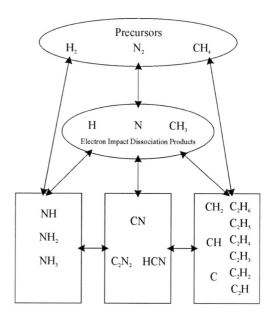

Figure 5.37: Schematic diagram
of the reaction kinetics in H_2-N_2-
Ar-CH_4 discharges.

A basic reaction scheme for a discharges containing the gas mixture of interest here is presented in Figure 5.37. In principle the 22 species included in the model can be gathered into five different groups. First, there are the precursor molecules and their electron impact dissociation products. These eventually form ammonia, hydro cyanide and methane or ethane by thermal chemical reactions in the discharge volume. Species of the cyanide group are mainly formed by the interaction of hydrocarbons with nitrogen or with species of the ammonia-producing group. In contrast, the production of ammonia or hydrocarbons was found experimentally to be relatively independent of the hydrocarbon chemistry. This fact was taken into account in the reaction kinetic model and in the basic reaction scheme.

5.6   Spectroscopic Investigations of the CN Free Radical

The CN free radical has played a central role in the development of molecular spectroscopy. Its well-known red and violet electronic band systems are ubiquitous in flames and discharges containing hydrogen and carbonaceous molecules. As a result the electronic spectra of the radical have been comprehensively investigated culminating in the recent results of Rehfuss et al. [191] and Prasad et al. [192] who used high resolution FT spectroscopy to measure CN spectra in jet sources. The radical has been equally important in microwave spectroscopy and rotational spectra of different vibrational states have been measured for both $^{12}C^{14}N$ [193] and $^{13}C^{14}N$ [194]. Both rotational [195] and electronic spectra [196] have been detected in astronomical objects such as interstellar clouds and stellar atmospheres.

In contrast, rather less effort has been expended on studying its vibration-rotation spectra, although they are being used increasingly for kinetic and dynamic investigations involving CN [144,197]. The rotationally resolved fundamental band was first measured in the laboratory at modest resolution (0.2...0.35 cm^{-1}) by Treffers [198]. He detected three lines using a King furnace as a source. Later, ten rotational lines of the fundamental band were measured in a glow discharge at Doppler limited resolution using infrared diode laser absorption spectroscopy [199]. A much wider region of the infrared was recorded by Davis et al. [200] using FTIR at 0.01 cm^{-1} resolution to measure sequences from (1-0) to (4-3) and from (2-0) to (4-2). The radical was again produced in a King furnace and spectral subtraction of the intense CO spectrum also present was required to uncover the CN transitions.

This chapter describes the measurement and analysis of more extensive diode laser spectra of two isotopic forms of CN in its ground state ($X^2\Sigma^+$). These new measurements were made for two reasons. Firstly, the measurement accuracy of diode laser spectroscopy has increased substantially since the earlier measurements, which were made more than twenty years ago [199]. The main reason being that etalon, calibration and CN spectra can now be recorded simultaneously rather than sequentially as earlier. A measurement accuracy as high as 0.0004 cm^{-1} can now be obtained, about a factor of ten better than the former measurements. Secondly, diode laser spectroscopy is becoming increasingly important for kinetic and dynamic studies of CN. When the radical is generated in plasmas, CN lines are often overlapped by other species, particularly CO transitions, and missassignments may occur. Hence precise CN line positions are essential. This chapter reports the measurement of CN ($X^2\Sigma^+$) fundamental band spectra recorded in a methane-nitrogen-hydrogen plasma and, for the first time, measurements of the $^{13}C^{14}N$ fundamental.

5.6.1   Experimental

A schematic diagram of the plasma reactor and diode laser spectrometer is shown in Figure 5.38. This type of planar microwave reactor has been used extensively for plasma chemical described in the preceding chapters. It was described in detail in chapter 5.1 to 5.4. In contrast to that modified discharge conditions, i.e. a lower pressure of 0.3 mbar and higher power of 3 kW, were used. For the spectroscopic investigations the experimental arrangement was extended by two additional beams, one passing a gas cell filled with the reference gas and one through an etalon with a fringe spacing range of 0.01 cm^{-1} only. Prior to measurements on $^{13}C^{14}N$, which required the use of $^{13}CH_4$, the entire reactor was cleaned of the C-N polymeric deposits characteristic of CH_4 / N_2 plasmas; it was then operated in a sealed-off mode. Spectra were recorded simultaneously with calibration gas lines (carbonyl sulphide [115,117] and carbonyl fluoride [117]) and several measurements made of each CN line. The accuracy of measurement was about 0.0005 cm^{-1}. For an estimated absolute concentration value of CN the line strength of the P(7) line quoted by Balla and Pasternack [142] was used.

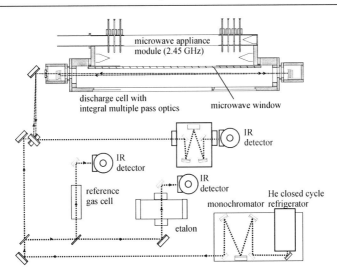

Figure 5.38: Experimental arrangement of the planar microwave plasma reactor (side view) with White cell multiple pass optical arrangement and tunable diode laser infrared source. The laser beam path is indicated by dotted lines.

5.6.2   Results

The behaviour of the CN bearing species in the plasma for a fixed amount of added methane and varying proportions of nitrogen and hydrogen are shown Figure 5.39. While the HCN concentration rises rapidly as the proportion of N_2 reaches 25 % and then is approximately constant, both CN and C_2N_2 rise, in contrast to that, steadily up to the maximum proportion of nitrogen (Figure 5.39). The main product of the plasma reactions is HCN, which has a concentration several orders of magnitude higher than CN or C_2N_2. Figure 5.40 and Figure 5.41 demonstrate how varying the total pressure and the applied plasma power affects the concentrations of CN and HCN. While both CN and HCN concentrations increase with power the two species show a contrary behaviour with the pressure. The CN radical concentration was estimated to be in the range of 10^{10} - 10^{11} molecules cm^{-3}.

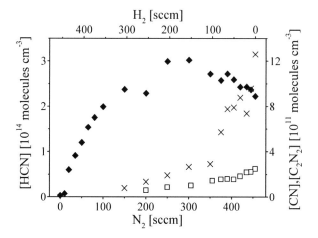

Figure 5.39: Variation of CN molecular species concentrations as a function of the proportion of H_2 to N_2. ($\blacklozenge$ - HCN, $\times$ - C_2N_2, p = 1.5 mbar, P = 1.5 kW, $N_2 + H_2 = 455$ sccm, Ar = 60 sccm, $CH_4 = 40$ sccm) ($\square$ - CN, p = 1.1 mbar, P = 3 kW, $N_2 + H_2 = 455$ sccm, Ar = 60 sccm, $CH_4 = 40$ sccm)

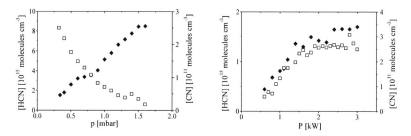

Figure 5.40: Concentrations of HCN ($\blacklozenge$) and CN ($\square$) as a function of total pressure (P = 3 kW, $N_2 : Ar : CH_4$ = 1 : 1 : 1)

Figure 5.41: Concentrations of HCN ($\blacklozenge$) and CN ($\square$) as a function of power (p = 0.33 mbar, $N_2 : Ar : CH_4$ = 1 : 1 : 1)

Searches for the infrared spectra of CN were straightforward using the available spectroscopic constants to predict rotational line positions. In most cases lines were free from possible interfering lines from other plasma species such as CO. Lines which are partly overlapped were given reduced weight in the fits. Figure 5.42 is a composite of several lines from $^{12}C^{14}N$ and $^{13}C^{14}N$ fundamental bands. In some cases further structure in the lines, such as spin-rotation splitting, is resolved. Whether this structure can be resolved depends in part on the tuneability of the diode. The minimum current step of the laser current is 10 μA and for a slow tuning diode (optimum case) this corresponds to a minimum wavenumber step of $\approx 5 \times 10^{-4}$ cm^{-1}. In principle this is adequate to resolve the ≈ 0.007 cm^{-1} spin-rotation splitting. A second factor, which may prevent the resolution of fine or hyperfine structure, is the temperature of the plasma. It was found that when the input power was increased the line width also increased corresponding to a higher temperature and greater Doppler linewidth.

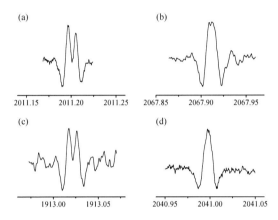

Figure 5.42: Representative absorption lines from the fundamental band of CN: (a) P(8) in ^{12}CN; (b) R(6) in ^{12}CN; P(22) in ^{13}CN; R(11) in ^{13}CN

Table 5.6 contains the measured lines from both isotopic species. The results of fitting to the standard formula for a vibration-rotation spectrum are given in Table 5.7. Although splitting of the lines was observed in several cases this was ignored for fitting purposes and the line position given in Table 5.6 is the average position of any resolved splitting. The data were fitted to five parameters: the band origin, and the upper and lower B_v and D_v values. Although J values up to 25 were measured, it was found to be unnecessary to include the sextic distortion constants in the fit (H$\approx 5 \times 10^{-12}$ cm^{-1}).

J'	J''	Measured Position	10^4(obs-calc)	weight
		$^{12}C^{14}N$		
14	15	1982.1167	3	1.0
13	14	1986.3699	-1	1.0
11	12	1994.7792	-3	1.0
9	10	2003.0603	34	0.1
8	9	2007.1451	-7	1.0
7	8	2011.2015	3	1.0
6	7	2015.2232	3	1.0
5	6	2019.2110	1	1.0
6	5	2064.3760	-5	1.0
7	6	2067.9118	4	1.0
8	7	2071.4043	-61	0.1
9	8	2074.8737	3	1.0
10	9	2078.3000	-1	1.0
11	10	2081.6915	10	1.0
12	11	2085.0414	-29	0.1
13	12	2088.3612	-3	1.0
15	14	2094.8852	0	1.0
17	16	2101.2604	-1	1.0
		$^{13}C^{14}N$		
23	24	1904.3740	0	1.0
22	23	1908.7118	-10	1.0
21	22	1913.0226	5	1.0
20	21	1917.3031	14	1.0
15	16	1938.2495	-12	1.0
7	8	1970.1781	1	1.0
6	7	1974.0287	4	1.0
5	6	1977.8473	4	1.0
4	5	1981.6339	2	1.0
3	4	1985.3880	-4	1.0
5	4	2017.7180	-2	1.0
6	5	2021.1474	21	0.1
10	9	2034.5164	-2	1.0
11	10	2037.7761	16	0.1
12	11	2040.9977	-4	1.0
15	14	2050.4623	5	1.0
16	15	2053.5468	-2	1.0
17	16	2056.5971	0	1.0
20	19	2065.5360	0	1.0
23	22	2074.1551	2	1.0
24	23	2076.9559	-2	1.0
26	25	2082.4583	85	0.1

Table 5.6: $^{12}C^{14}N$ and $^{13}C^{14}N$ ($X^2\Sigma^+$) fundamental band lines (in cm^{-1})

	$^{12}C^{14}N$	$^{13}C^{14}N$
B_0	1.891073(27)	1.812670(14)
D_0	$6.40(8) \times 10^{-6}$	$5.89(2) \times 10^{-6}$
B_1	1.873646(25)	1.796349(12)
D_1	$6.39(8) \times 10^{-6}$	$5.90(1) \times 10^{-6}$
v_0	2042.42104(84)	2000.08470(30)
s. d.	0.00076	0.00057

Table 5.7: Fitted parameters (in cm^{-1}) for $^{12}C^{14}N$ and $^{13}C^{14}N$ ($X^2\Sigma^+$)
(one standard deviation given in parenthesis)

5.6.3   Discussion

The source of inverted populations of HCN in the far infrared discharge laser has been a subject for much discussion. Pichamuthu [201] and others have suggested that reaction of vibrational excited $X^2\Sigma^+$ CN with H_2 is the source of the excited HCN. The conditions in the plasma investigated here are not greatly different from those used in the FIR HCN laser. It seems unlikely then, given the relative concentrations of CN and HCN (Figure 5.39) that CN is the precursor of HCN. Furthermore, Skatrud and DeLucia [202] have proposed a detailed model involving vibrationally excited nitrogen as the excitation precursor of inverted HCN. The reaction(s) leading to the formation of HCN in its ground state are not specified in their mechanism. Hence at the present time the reactions responsible for the large amounts of HCN produced in nitrogen and methane plasmas are uncertain.

The origin of the fundamental band of the $^{12}C^{14}N$ species given in Table 5.7, 2042.42104(84) cm^{-1}, is the most precise reported so far and agrees with other determinations within their 3σ values (given in parenthesis). For example,

2042.4183(39) cm^{-1} [200],   2042.4226(21) cm^{-1} [203],   2042.406(52) cm^{-1} [192].

The most directly comparable with the diode laser determination is that due to Davis et al. [198] which also derives from the vibration-rotation spectra.

Much less data is available for $^{13}C^{14}N$. However, the value of v_0 lies within the 3σ uncertainty quoted by Hosinsky et al. [204] who obtained 2000.103(24) cm^{-1}. A rather closer value is obtained using Cerny et al.'s [203] data (quoted in reference [204]) namely 2000.092 cm^{-1} (no uncertainties given).

The B values for the $^{12}C^{14}N$ species (Table 5.7) are in very close agreement with the results of the FTIR study [200]. The values derived from the equilibrium parameters of Davis et al [200] are B_0= 1.891076(9) and B_1= 1.873657(9) cm^{-1} (1σ). There is less satisfactory agreement with the results of the analysis of the red system [204] of $^{13}C^{14}N$ which yields B_0= 1.81253(6) and B_1= 1.79617(6) cm^{-1} but better agreement with the values calculated from Cerny et al. [203] (also quoted in reference [204], B_e= 1.82082 cm^{-1}, α_e= 0.0163 cm^{-1}), namely B_0= 1.81267 and B_1= 1.79637 cm^{-1}. Hosinsky et al also report modified parameters from analysis of the (0,0) band of the red system by Brault and Engleman (unpublished work). The value of B_0 is given as 1.81268(5) cm^{-1}, in good agreement with the value in Table 5.7.

Whilst the agreement between the diode laser and optical results for the rotational constants are highly satisfactory, a more rigorous test is against the microwave values. The four B values reported in Table 5.7 all agree with the microwave-determined values to about 1σ or better. For $^{12}C^{14}N$ the microwave values are B_0=1.891091 and B_1=1.873667 cm^{-1}, both about 2×10^{-5} cm^{-1} higher than those in Table 5.7. Nevertheless, the vibration-rotation constants (α_e) from microwave and diode laser measurements are in near exact agreement. The $^{13}C^{14}N$ microwave values are B_0=1.812692 and B_1=1.796344 cm^{-1}, in rather better agreement with the results in Table 5.7. However, for the latter there is a significantly smaller standard deviation for the fit of the diode laser data. The consequence of not resolving the fine and hyperfine structure completely can be seen in the experimental line-widths such as in the spectra in Figure 5.42. For example, the calculated width [203] of the line in 6(d) assuming a Doppler profile and temperature of 1000 K for the plasma is 0.0087 cm^{-1} (minimum to minimum of the second derivative line profile). The measured line-width is 0.021 cm^{-1}. Without the availability of sub-Doppler spectra, or at least of spectra with reduced Doppler width as in a supersonic jet source, the method of incorporating these effects would be to calculate the fine / hyperfine patterns individually for each infrared line and convolute them with a Gaussian line profile. In the absence of such calculations the parameters listed in Table 5.7 should be regarded as effective ones. Lastly, to facilitate searches for other lines in the fundamental bands Table 5.8 and Table 5.9 list all the line positions up to J= 25 calculated using the constants in Table 5.7. Comparison with the earlier diode laser measurements [199] for $^{12}C^{14}N$ shows very satisfactory agreement.

P25	1937.8170	R0	2046.1683
P24	1942.3886	R1	2049.8806
P23	1946.9293	R2	2053.5577
P22	1951.4386	R3	2057.1995
P21	1955.9166	R4	2060.8058
P20	1960.3631	R5	2064.3765
P19	1964.7778	R6	2067.9114
P18	1969.1607	R7	2071.4105
P17	1973.5115	R8	2074.8734
P16	1977.8301	R9	2078.3001
P15	1982.1164	R10	2081.6905
P14	1986.3702	R11	2085.0443
P13	1990.5912	R12	2088.3615
P12	1994.7795	R13	2091.6418
P11	1998.9348	R14	2094.8852
P10	2003.0569	R15	2098.0915
P9	2007.1458	R16	2101.2605
P8	2011.2012	R17	2104.3921
P7	2015.2229	R18	2107.4861
P6	2019.2109	R19	2110.5424
P5	2023.1650	R20	2113.5609
P4	2027.0850	R21	2116.5413
P3	2030.9707	R22	2119.4836
P2	2034.8221	R23	2122.3876
P1	2038.6389	R24	2125.2532
		R25	2128.0802

Table 5.8: Calculated $^{12}C^{14}N$ ($X^2\Sigma^+$) fundamental band lines (in cm^{-1})

P25	1900.0059	R0	2003.6774
P24	1904.3741	R1	2007.2372
P23	1908.7128	R2	2010.7641
P22	1913.0221	R3	2014.2578
P21	1917.3018	R4	2017.7183
P20	1921.5516	R5	2021.1453
P19	1925.7716	R6	2024.5388
P18	1929.9615	R7	2027.8986
P17	1934.1213	R8	2031.2246
P16	1938.2507	R9	2034.5166
P15	1942.3497	R10	2037.7745
P14	1946.4181	R11	2040.9981
P13	1950.4557	R12	2044.1873
P12	1954.4625	R13	2047.3419
P11	1958.4382	R14	2050.4619
P10	1962.3829	R15	2053.5470
P9	1966.2962	R16	2056.5971
P8	1970.1780	R17	2059.6121
P7	1974.0284	R18	2062.5918
P6	1977.8469	R19	2065.5360
P5	1981.6337	R20	2068.4448
P4	1985.3884	R21	2071.3178
P3	1989.1111	R22	2074.1549
P2	1992.8014	R23	2076.9561
P1	1996.4593	R24	2079.7211
		R25	2082.4498

Table 5.9: Calculated $^{13}C^{14}N$ $(X^2\Sigma^+)$ fundamental band lines (in cm^{-1})

6   Investigations in DC Discharges

6.1   Investigations of the Carbon Dioxide Conversion Chemistry in a Low Pressure Glow Discharge

6.1.1   Introduction

Within the last decades, the kinetics of the CO_2 chemistry has been in the centre of interest of several studies. Absorption spectroscopic methods, using diode lasers in the near [207-210] and mid infrared region [211-213], have been applied to monitor concentrations of CO and CO_2 in the combustion of burner flames under atmospheric and low-pressure conditions. These measurements were performed with temporal resolutions between some milliseconds [210] and some minutes [213]. Morvova [213] studied the development with time of the conversion chemistry in dc corona discharges of CO-air and CO_2-air mixtures using a FTIR absorption spectrometer. Recently Kylian et al. [215] investigated the time evolution and the steady state of the CO_2 decomposition in DC glow discharges in sealed-off CO_2 laser mixtures by means of time resolved optical emission spectroscopy.

The work presented in this chapter is based on publications of Rutscher et al. [216], Lucke et al. [217], Sonnenfeld et al. [218] and Gundermann et al. [219]. Sonnenfeld et al. [218] monitored plasma chemical processes in different low-pressure non-thermal gas discharge systems (dc positive column, hf) containing CO_2 using gas chromatography. Their experimental investigations were focused on the establishment of so-called chemical quasi-equilibrium states in two different reactor types: the flow reactor (divided in an active zone and a passive zone) and the closed system (temporally combination of plasma phase and relaxation phase). The gas analysis was performed by extracting gas samples in the temporal afterglow of the discharge. In addition they discussed a basic kinetic model of the $CO_2 / CO / O_2$ conversion chemistry. Gundermann et al. [219] determined the electron density in $CO_2 / CO / O_2$ mixtures in the dc positive column under similar plasma conditions, studied here, using microwave diagnostic methods.

In this chapter the plasma chemical decomposition of CO_2 is studied in a further kind of closed reactors, characterized by a stationary active zone, separated in space and surrounded by a stationary afterglow. The investigations were focused on tunable diode laser absorption spectroscopic studies in the mid infrared region that had been performed in the extended stationary afterglow of a closed dc glow discharge of CO_2. For the first time absolute concentrations of CO and CO_2 were measured time and spatially resolved in such a low pressure discharge system to monitor the temporal development of the chemical conversion process in the CO_2 / CO / O_2 system.

6.1.2   Experimental

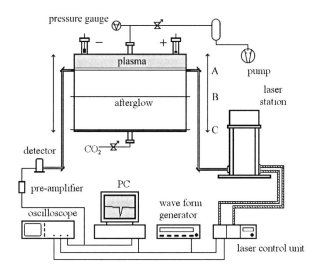

Figure 6.1: Experimental set-up with measurement positions (A – afterglow near the discharge, B – center of the afterglow, C – end of the afterglow)

The experimental set-up of the used absorption spectroscopy system and the discharge system is shown in Figure 6.1.

The small plasma zone and the long extended stationary afterglow were located in a closed stainless steel vacuum chamber of 48 cm x 30 cm x 6 cm. Figure 6.2 shows a schematic side view of the vessel. The cylindrical discharge electrodes (7) are made from aluminium with a diameter of 1 cm. The current of the dc glow discharge varied between 2 – 30 mA. The discharge area, in particular the positive column was located in a glass tube (6), which had a diameter of 3 cm and a length of 48 cm. This glass discharge tube was connected to the two electrodes and located at one wall of the chamber. The tube had a small lateral slit (5) of 1 mm width over the whole length at the tube wall to the open chamber. The afterglow filled the rest of the discharge vessel. The centre supporting part (3) was necessary to fix the windows to the metallic vessel.

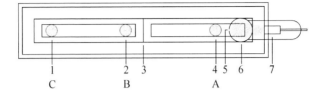

Figure 6.2: Schematic side view of the discharge vessel (1 – measurement position C at the end of the afterglow 22 cm from the slit, 2 – measurement position B in the center of the afterglow 12 cm from the slit, 3 – center supporting part for the KBr windows, 4 – measurement position A in the afterglow near the discharge 1 cm from the slit, 5 – side view of the 1 mm slit in the discharge tube, 6 – discharge tube, 7 - cylindrical aluminum discharge electrodes)

Before generating a plasma the discharge the vessel was filled with CO_2 at an initial pressure of 50 Pa. Then the volume was closed, the plasma was ignited and the temporal development of the molecular species concentration was observed. The pressure in the chamber was found to increase related to the CO_2 conversion to CO. After stopping the discharge and the measurement the system was evacuated to a pressure of less than 0.01 Pa before refilling. Measurements were repeated five times and averaged.

For the time and spatially resolved tunable infrared diode laser absorption spectroscopic measurements in the afterglow of the discharge a TDLAS system built by Muetek Infrared Laser Systems with two infrared lead-salt diode lasers was used. The diode lasers were mounted in a laser station with a helium closed cycle cryostat (see Figure 6.1). The temperature of the lasers could be controlled at milli-Kelvin precision in the range between 25 K and 80 K. The infrared laser beam passed a grating monochromator that servers as a mode filter (not shown in Figure 6.1) before entering the discharge chamber via KBr windows located on two sides of the vessel. The absorption length of the single pass through the discharge vessel was 48 cm. The laser beam position in the plasma reactor could be adjusted using a set of moveable optical mirrors. Measurements were performed at three different positions, (A) 1 cm away from the exit slit of the discharge tube, (B) in the center region of the afterglow in 12 cm distance to the slit and (C) at the end of the afterglow 22 cm from the slit of the discharge tube. The laser signal was recorded using an infrared detector mounted in a liquid nitrogen dewar. The absorption signal of the detector was amplified and transferred to a PC system. In addition, the spectra could be visualized using a Tektronix TDS 714L digital oscilloscope.

The diode laser was driven with a rectangular pulse of a frequency of 2.5 kHz provided by a HP 33120A waveform generator. The data acquisition was performed using a SPECTRUM 12 bit PAD1232 A/D-converter transient recorder board in the PC system with a sample frequency of 30 MHz. Data processing was realized by the software DENSIDET programmed according to the requirements. DENSIDET reads out the memory of the transient recorder card and averages several diode laser scans to improve the signal to noise ratio. Additionally the program performed (i) the determination of the values of the background intensity without absorption, I_0, (ii) the determination of the transmitted laser intensity I and (iii) the on line calculation of the intensity ratio I / I_0. The intensity ratio was displayed versus time and saved automatically. A time resolution of three seconds was chosen for the measurements of the temporal development of the CO and CO_2 concentration, lasting five minutes.

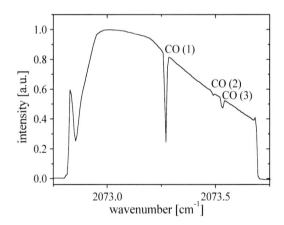

Figure 6.3: Example of a TDL absorption spectrum with three CO lines in the 2073 cm^{-1} wavenumber region (see table 1) measured in 1 cm distance to the discharge tube (position A, pulsed diode, f = 2.5 kHz)

The absorption lines of CO and CO_2 used for concentration measurements and line identification, their absolute positions and line strengths are listed in Table 6.1. A representative direct tunable diode laser absorption spectrum measured nearby the slit of the discharge tube with three lines due to CO is shown in Figure 6.3. For the measurements of the CO concentration line (1) was used. In this wavenumber region at plasma on conditions it was possible to observe a hot band of excited CO in the afterglow of the discharge too (not shown in Figure 6.3).

Molecule line		Line position (cm^{-1})	Line strength (cm / molecule)	Refs.	Remarks
CO	(1)	2073.2646	9.087×10^{-20}	[116]	used for measurements
	(2)	2073.4892	6.640×10^{-22}	[116]	used for identification
	(3)	2073.5252	3.990×10^{-21}	[116]	used for identification
CO_2		617.1895	2.302×10^{-21}	[116]	used for measurements

Table 6.1: Molecular absorption lines used for line identification and concentration measurements

The identification of the absorption lines and the measurement of their absolute positions was carried out using the wavenumber calibration of the grating monochromator of the spectrometer, well documented reference gas spectra, gas cells containing reference gasses with known pressures and purity and an etalon of known free spectral range for wavenumber interpolation [114,116,118]. With the DENSIDET data output of the intensity ratio I / I_0 the species density was easy to calculate using Lambert-Beer's law. Based on the profile and intensity analysis of the CO absorption lines with and without a discharge a gas temperature of about 350 ± 50 K could be estimated. This observation is in accordance with the rather low power input of about 30 W in maximum into the discharge, leading to an only slight heating of the discharge vessel of about 10...20 K above room temperature.

6.1.3   Results and Discussion

6.1.3.1   Investigation of the Temporal Development of CO and CO_2

Figure 6.4 shows the measured temporal development of the CO concentration during the CO_2 conversion process at the different positions in the afterglow of the DC glow discharge with a discharge current of 30 mA. For clarity reasons not every measured point is shown in this figure.

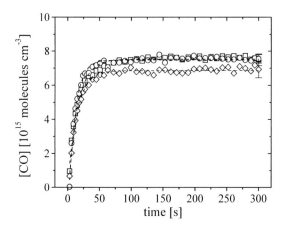

Figure 6.4: Time dependence of the CO concentration for the three different positions in the afterglow (not every measured point shown for clarity, I = 30 mA, □ - near the discharge (A), ○ - center of the afterglow (B), ◇ - end of the afterglow (C))

As one can see the temporal development and the equilibrium value of the concentration of CO did not depend on the measuring position in the afterglow of the discharge for a given pressure and discharge current within the accuracy of measurements. The equilibrium value of the product concentration was reached at the same time near the plasma zone, in the middle of the afterglow as well as at the end of the afterglow. The slightly reduced CO equilibrium concentration found at the end of the afterglow was still within the error bars. This result can be explained under the given experimental conditions by the relative large diffusion coefficient of about 50 cm^2 s^{-1} for the species transport along the chamber. A change in the species concentration in the active plasma zone led to a steady state in some tenth of a second in the whole chamber. For this reason each small concentration gradient vanished in the afterglow part of the chamber faster than it could be detected with a time resolution of three seconds used for the measurements at the various positions.

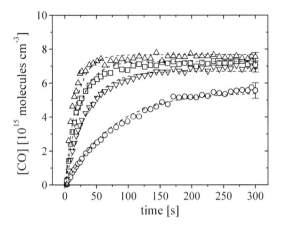

Figure 6.5: Comparison of the measured temporal developments of the CO concentration (symbols) with results of a numerical modelling of the chemical conversion kinetics (dashed lines) for four different discharge currents (not every measured point shown for clarity, center position (B), $\triangle$ - 30 mA, $\square$ - 20 mA, ∇ - 10 mA, $\bigcirc$ - 2 mA)

In Figure 6.5 the time dependence of the CO concentration in the centre of the afterglow is shown for four discharge currents. Again for better view not all measured values are visualized. In agreement with former investigations [216-218] the formation of quasi-equilibrium CO concentration values, depending insignificantly on the current, was found. The graphs illustrate as well the strong dependence of the formation time of the stationary species composition in the reactor on the discharge current. With increasing current, the number of per time unit produced CO molecules in the first seconds of the discharge grew up. This led to a reduced time necessary to reach the equilibrium product composition.

The temporal developments of the CO concentration could be fitted for the various currents by exponential functions (not shown in Figure 6.5):

$$[n] = [n]_E - A_1 \, exp(-t \, / \, t_1).$$

These fits gave values for the equilibrium concentrations $[n]_E$ and characteristic time constants t_1 which described the time necessary for the formation of a stationary product composition in the afterglow. Due to the fact that the starting concentration of CO is zero, $[n]_E$ and A_1 are equal. Figure 6.6 shows the small linear rise of the equilibrium concentration value $[n]_E$ and the strong exponential decrease of the formation time constant t_1 with increasing discharge current for measurements of the temporal development of CO in the centre of the afterglow. The numerical results for the equilibrium concentration $[n]_E$ and the formation time constant t_1 are given in Table 6.2. With the density of the CO_2 before starting the plasma of about 13×10^{15} molecules cm^{-3} and the equilibrium concentration values one can calculate a final conversion ratio. This is listed in Table 6.2, too. The increase of the equilibrium concentration values of CO and of the related final conversion ratio was mainly caused by the larger energy input into the discharge. This influences the chemical balance established in the plasma and in the spatial afterglow.

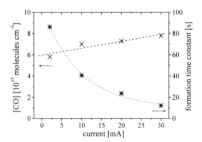

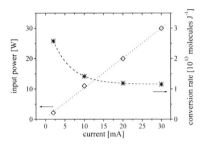

Figure 6.6: Results of exponential fitting of the measured temporal developments of the CO concentrations as a function of the discharge current (center position (B), $\times$ – equilibrium concentration value (left axis), $\ast$ – formation time constant (right axis))

Figure 6.7: Input power and estimated conversion rate as a function of the discharge current (center position (B), $\diamond$ - input power (left axis), $\ast$ – conversion rate (right axis)

Current (mA)	Molecule	Equilibrium Concentration (molec cm^{-3})	Formation Time Constant (s)	Final Conversion Ratio (%)	Conversion Rate (molec J^{-1})	Electron density (cm^{-3})
2	CO	5.8×10^{15}	86.3	43.6	2.58×10^{13}	2.6×10^9
10	CO	7.0×10^{15}	40.7	52.6	1.42×10^{13}	7.0×10^9
20	CO	7.3×10^{15}	23.7	54.9	1.19×10^{13}	1.1×10^{10}
	CO_2	4.5×10^{15}	24.1	66.2	1.38×10^{13}	
30	CO	7.8×10^{15}	12.4	58.6	1.16×10^{13}	1.5×10^{10}

Table 6.2: Equilibrium concentrations of CO and CO_2, formation time constants and conversion ratios and rates derived by fitting the measured temporal developments in the center of the afterglow and electron densities used in the model for various discharge currents. The electron density for 2 mA was taken from [219]. The values for the higher currents are linearly approximated

In the first seconds of running the discharge the concentration of CO increased almost linearly with time. Depending on the power consumption of the plasma one could estimate a value for a conversion rate in the CO_2 / CO discharge system, i.e. the number of from CO_2 to CO converted molecules per energy value. The results of this estimation are also listed in Table 6.2. In Figure 6.7 the conversion rate is shown together with the values of the power input into the discharge. The input power on the two electrodes increased linearly with the current. The conversion rate was for currents higher than 10 mA estimated to 1.2×10^{13} molecules J^{-1}. For lower currents (such as 2 mA) the estimated conversion rate was twice bigger than for currents of $10 - 30$ mA. This means on the one hand that the energy efficiency of the conversion process for low currents was improved. On the other hand, since the input power was much lower it needed a longer time to establish a chemical equilibrium in the conversion process. It should be noted, that at a low discharge current of 2 mA the discharge characteristics differed considerably from conditions at all higher currents. Changed plasma properties, in particular a changed electron energy distribution function, could be considered as the reason.

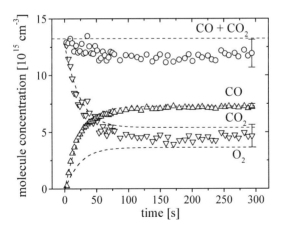

Figure 6.8: Comparison of the measured temporal developments of the CO and the CO_2 concentrations and its sum (symbols) with results of a numerical modelling of the chemical conversion kinetics (dashed lines) (not every measured point shown for clarity, center position (B), I = 20 mA, $\nabla - CO_2$, $\triangle - CO$, $O - CO + CO_2$)

A comparison of the time dependences of the CO and the CO_2 concentration in the centre of the afterglow is given in Figure 6.8 for a discharge current of 20 mA. It was found that the concentration of CO_2 decreased in the same time scale of about 24 seconds as the density of CO increased. The mass balance could be considered as nearly fulfilled within the accuracy of the measurements. For this reason, the gas composition could be determined by knowledge either of the CO or the CO_2 concentration, only.

This quasi-equilibrium of the two species concentrations is of course determined by the type of plasma source, in particular if it is unpulsed or pulsed dc or frequency modulated, the discharge parameters like pressure or discharge power, and the properties of the surrounding walls. It should be possible to move if necessary the chemical equilibrium to higher conversion percentages e.g. by changing the wall material or temperature.

6.1.3.2   Modelling

For further insight to the kinetics in the chemical conversion the measured temporal developments of the CO and the CO_2 concentrations in the centre of the afterglow of the discharge was modelled. For this purpose, FACSIMILE was used. This program package was already described in chapter 5.5.3.

6.1.3.2.1   The Basic Model

A rather simple reaction scheme of 10 electron impact dissociation, wall induced recombination and thermally activated reactions in the volume was used to model the kinetics of the chemical conversion process of CO_2 to CO. These reactions and their specific rate coefficients at a temperature of 350 K are given in Table 6.3.

No.	Reaction	Rate Equation $((cm^3 / molecule)^{n-1} s^{-1})$	Rate Coefficient $(cm^3 \, molecule^{-1} \, s^{-1})$
1	$CO_2 + e^- = CO + O + e^-$		$2.0 \times 10^{-12} \, n_e$
2	$CO + e^- = C + O + e^-$		$5.0 \times 10^{-13} \, n_e$
3	$O_2 + e^- = O + O + e^-$		$2.0 \times 10^{-11} \, n_e$
4	$CO + O + Wall = CO_2 + Wall$		2.0×10^{-15}
5	$O + O + Wall = O_2 + Wall$		1.1×10^{-11}
6	$CO + O + M = CO_2 + M$	$2.76 \times 10^{-32} \exp(-1764/T)$	2.38×10^{-18}
7	$O + O + M = O_2 + M$	$2.7 \times 10^{-32} \, T^{-0.5}$	1.92×10^{-17}
8	$C + CO_2 = CO + CO$	$1.74 \times 10^{-13} \, T^{0.5} \exp(-1703/T)$	2.32×10^{-14}
9	$C + O + M = CO + M$	$8.27 \times 10^{-32} \, T^{0.5}$	5.88×10^{-17}
10	$C + O_2 = CO + O$	$8.8 \times 10^{-13} \, T^{0.5} \exp(-1661/T)$	1.43×10^{-13}

Table 6.3: Rate equations and rate coefficients of electron impact dissociation, wall recombination and thermally activated chemical reactions used to model the chemical conversion of CO_2 (n – number of reacting species, n_e – electron density) [218,219]

For the modelling of the kinetics of the conversion process only the most important reactions that decompose or produce CO_2, CO, C, O and O_2 were taken into account. The rate coefficients of electron impact dissociation reactions of CO_2 in discharges and wall induced recombination of species containing carbon were adapted related to other references like Sonnenfeld et al. [218] or Mechold et al. [220]. The rate equations for the thermally activated volume reactions were taken from Sonnenfeld et al. [218], where a similar model is described. Sonnenfeld et al. [218] assumed that due to the low ionisation degree and the low discharge pressure reactions of exited neutrals and charged particles are of minor importance. These reactions were not included in this model. Other thermally activated volume reactions with small rate coefficients were neglected in the model too.

The value of the rate coefficient for the electron impact dissociation of CO_2 played a key role in the model. It influenced to a less extent the formation time of the conversion equilibrium but more seriously, it influenced the equilibrium concentration values of CO_2 and CO. Nevertheless, the rate coefficients used in this work are smaller compared to Sonnenfeld et al. [218]. The recombination to CO_2 on the wall seemed to be important for the formation time of the conversion equilibrium. The results of the measurements of the temporal development of the CO concentration at various positions in the afterglow showed no significant dependence on the distance to the active discharge (see Figure 6.4). Therefore, the discharge vessel with the tube containing the plasma and the wide cavity filled with the stationary afterglow where the measurements were made was treated in the model like one big cell. No transportation and no diffusion effects in the chamber were taken into account since they had been proofed to have no effect for the used time resolution (see above). Based on experiments an estimated gas temperature of 350 K was used for the modelling of the chemistry. For the dissociation processes initiated by electron impact an electron density of 1.1×10^{10} cm^{-3} was assumed for a discharge current of 20 mA. This value is in between the values used in ref. [218] and ref. [220]. The electron density of 2.6×10^9 cm^{-3} at 2 mA is in well agreement with the values measured by Gundermann et al. [219], where electron densities at this pressure in discharge tubes with the same diameter under comparable conditions are reported. They found a value of 28.17 V $Torr^{-1}$ cm^{-1} for the reduced electric field and an averaged kinetic electron temperature of 1.1 eV. For the starting concentration of CO_2 in the model the first measured CO_2 density of 1.3×10^{16} molecules cm^{-3} was used. In Figure 6.8 this value is shown as a dashed line to be compared with the summation of the measured concentrations of CO and CO_2. The same density was used for the collision partner M in the calculation of the rate coefficients. In addition, it was assumed that the concentrations of the species produced in the discharge like CO or O_2 were zero before starting the plasma.

6.1.3.2.2   Results of Model Calculations

The results of modelling the temporal development of the CO concentration (dashed lines) are presented for the four different discharge currents in Figure 6.5 together with the measured development (symbols). To match the experimental data only the electron density in the model had to be altered. The values for the four different currents are given in Table 6.2 too. The well agreement of the calculated temporal development and the measured behavior for 2 mA, were the electron density from Gundermann et al. [219] was used in the model, proofed the validity of the used simple reaction scheme.

In Figure 6.8 one can compare the modelled time dependence of the CO_2 and CO concentrations (dashed lines) with the measured behavior (symbols) in the center of the afterglow for a discharge current of 20 mA. In addition, the calculated value for the carbon content as the sum of the measured CO_2 and CO concentrations and the concentration behavior of the molecular oxygen given by the model are shown. The equilibrium concentration of the atomic oxygen calculated by the model was about 8.7×10^{12} molecules cm^{-3}, the result for the carbon was 6.2×10^{10} molecules cm^{-3}. Although various simplifications have been made for modelling the temporal concentration development the agreement of the experimental and modelling data is rather satisfying.

6.2   HCl Concentration Measurements in Pulsed H₂-Ar-N₂-TiCl₄ DC Plasmas

6.2.1   Introduction

This chapter describes TDLAS studies of the conversion of $TiCl_4$ into HCl in a pulsed H_2-Ar-N_2-$TiCl_4$ DC discharge. The objective of these investigations was to observe the amount of Cl converted from $TiCl_4$ to HCl in the gas phase by plasma chemical means. It was known that to high $TiCl_4$ contents result in certain Cl content in the deposited surface layer and a decreased wear resistance of the layer. The HCl concentration in the plasma was monitored as the added precursor concentration and the current density were varied, while maintaining constant discharge pressure and total flow rate.

6.2.2   Experimental

The experimental arrangement of the dc plasma reactor and the tunable diode laser system built by Muetek Infrared Laser Systems is shown in Figure 6.9.

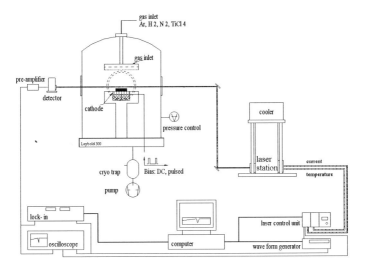

Figure 6.9: Scheme of the experimental set-up

The gas flows were measured on mass flow controllers and the gases mixed before entering the reactor. The gas was pumped out via a port in the bottom of the reactor. The gas mixture supplying the reactor consisted of 417 sccm H_2 + 89 sccm Ar + 77 sccm N_2 + 0.24...1.81 sccm $TiCl_4$. The total flow rate and pressure were kept constant at 583.2...584.8 sccm and 2 mbar, respectively. The pumping speed was adjusted with a butterfly valve to maintain a constant pressure in the plasma and the pressure measured on a capacitance pressure gauge. The discharge voltage was varied between 500 and 600 V leading to discharge currents between 60 mA and 120 mA (current densities: 0.6...1.15 mA cm^2).

The infrared diode laser beam from the TDL source assembly entered the plasma chamber via a KBr window. Its distance from the cathode was 3 cm. Before entering the plasma reactor, the diode laser beam first passed a mode selection monochromator (not shown). After leaving the reactor, again through a KBr window, for detection of the laser beam a HgCdTe detector was used.

The identification of lines and the measurement of their absolute line positions were performed using well-documented calibration spectra [116], reference gas cells containing methane and HCl and an etalon of given free spectral range. Because of the low absorption signal of the HCl molecule line at 3085.679 cm^{-1}, used for measurements, and to improve the signal to noise ratio, the concentrations were recorded using modulation of the diode laser current and phase sensitive detection at the second harmonic of the modulation frequency (derivative spectroscopy). The peak height of the second harmonic signal was calibrated by comparison with the signal of this line from the reference gas cell containing a known concentration of HCl, calculated from direct absorption measurements. The distance between the two KBr windows (70 cm) was used as absorption length. It was ensured that contaminations on plasma vessel walls had no influence on the measurements.

6.2.3   Results and Discussion

The concentration of HCl measured at $TiCl_4$ flows of up to 1.81 sccm by TDLAS with the discharge voltage as parameter is shown in Figure 6.10. The flow used in Figure 6.10 and Figure 6.11 for $TiCl_4$ was selected to represent typical flows in practical CVD reactors. Lower flows gave poorer signal-to-noise ratios for the used spectral line of HCl. In Figure 6.10 the broken line represents a maximum of HCl concentration assuming (i) a complete dissociation of the $TiCl_4$ precursor molecules and (ii) only the reaction of Cl atoms with hydrogen to produce HCl. Within the experimental errors this assumption, a nearly complete conversion of Cl into HCl, seems to be valid up to $TiCl_4$ admixtures of about

1 sccm. For these relatively small precursor flow rates no dependence on the discharge voltage could be found. Higher TiCl$_4$ admixtures lead to HCl concentrations of about 4...6 x 10^{14} molecules cm^{-3}, significantly depending on the discharge voltage.

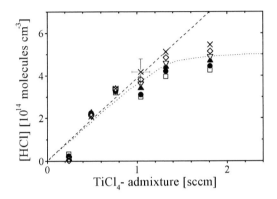

Figure 6.10: HCl concentration as a function of the TiCl$_4$-admixture for various discharge voltages (□ - 500 V, ● - 520 V, ▲ - 540 V, ▽ - 560 V, ◇ - 580 V, ✕ - 600 V) in comparison with the estimated HCl-concentration for 100 % dissociation of TiCl$_4$ (dashed line). The dashed line illustrates a supposed dependence of the HCl-concentration on the TiCl$_4$-admixture. (gas mixture: 417 sccm H$_2$, 89 sccm Ar, 77 sccm N$_2$, 0.24...1.81 sccm TiCl$_4$, p = 2 mbar)

Figure 6.11 shows the concentration of HCl on the current density in the discharge region calculated from the measured discharge current and the area of the powered cathode. In accordance to Figure 6.10 up to TiCl$_4$ admixtures of about 1 sccm the HCl concentration depends only on the admixture not on the current density caused by a complete dissociation of the precursor molecule. At higher TiCl$_4$ admixtures, the concentration of HCl increases with the current density.

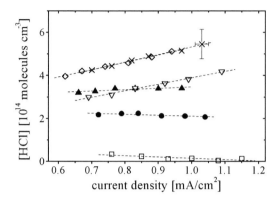

Figure 6.11: HCl concentration as a function of the current density for various TiCl₄-admixtures ($\square$ - 0.24 sccm (0.04 %), $\bullet$ - 0.48 sccm (0.08 %), $\blacktriangle$ - 0.76 sccm (0.13 %), $\triangledown$ - 1.04 sccm (0.18 %), $\diamondsuit$ - 1.32 sccm (0.23 %), $\times$ - 1.81 sccm (0.31 %)) (gas mixture: 417 sccm H_2, 89 sccm Ar, 77 sccm N_2, 0.24...1.81 sccm $TiCl_4$, p = 2 mbar, discharge voltage: 500...600 V)

The observation of HCl by diode laser spectroscopy in pulsed H_2-Ar-N_2 dc plasmas at small $TiCl_4$ admixtures has enabled to study the dissociation and conversion of the precursor molecule. From the results of measurements, $TiCl_4$ amounts of 1 sccm and average discharge voltages should be preferred for future optimised discharge experiments in these PACVD-reactor. Since for higher admixtures no complete conversion of chlorine into HCl was observed, for these flows it could be possible that with higher probability Cl is bound in the deposited surface layer.

Although this is the first time HCl concentration has been monitored in plasmas containing $TiCl_4$, it is clear that an extension of the present studies to other intermediates is desirable. In particular, other active plasma compounds at higher precursor admixtures should be investigated.

7   Accuracy and Limitations

The accuracy of measurement and in particular of species densities determined using TDLAS techniques in gases or gas discharges is limited, in principle, by the signal-to-noise ratio of the investigated absorption spectra. In addition, uncertainties of the necessary input data, like absorption length, line strength values or gas temperatures and pressure, used for the calculation of concentration results are important too. Since these problems are rather complex, only a rough overview is given here.

One of the main aspects for the quality, in particular the sensitivity and reliability, of the absorption measurement are the properties of the diode laser itself, in particular its spectral characteristics (single or multi-mode behaviour), stability and power output. In addition, the properties of the TDLAS system used, in particular the stability of the thermal control of the diode laser stations is of importance for the reliability and in particular for the reproducibility of measurement results. The quality of the optical arrangement, e.g. the spectral characteristics of the mode selector module or other laser power losses in the measuring path, the sensitivity and the noise level of the detectors and the noise level of the electronic data recording and handling system can influence the measured signal as well.

In the field of plasma diagnostics, the accuracy of measurements is often governed by the instabilities of the plasma itself. For this reason a compromise between sensitivity and time resolution for a specific measurement has to be made. Since millisecond time resolution was not needed in the present work, the signal was averaged for several seconds until a sufficient signal to noise ratio was reached. Even for the time resolved measurements reported above an accumulation time of 0.5-2 sec was sufficient for an appropriate signal. The noise produced by the experimental arrangement and the discharge was 5 % in maximum. Due to the averaging method it was therefore of minor importance.

The reliability of the calculated species concentrations from measured absorptions depends mainly on the accuracy of the molecular data and on experimental parameters, like total pressure, gas temperature or absorption length. The relative accuracy of line strength values of stable species is in the order of 10-20 % or even better. For free radicals and other unstable molecules a greater uncertainty of the line strengths has to be taken into account. For example, for CN it is $\approx$ 30-40 %.

Another parameter of importance is the absorption length. Stable species fill the complete reactor. Therefore, the absorption length is the distance between entrance and exit window and well known. Radicals mainly exist in the active zone of the discharge. For this reason, for these species the size of the visible plasma zone was used as absorption length. Since the border of the plasma is not sharp, an uncertainty of about 10 % can be estimated. The assumption of a gas temperature of 1000 K adds a systematic error of about 5 % in maximum. This was discussed in more detail above like the influence of errors of the total gas pressure.

In conclusion, absolute uncertainties in concentration measurements are approximately 10-15 % for the stable molecules and somewhat larger ($\geq$ 20 %) for the radicals i.e. CN and CH_3.

8   Summary and Conclusions

Tunable diode laser absorption spectroscopy (TDLAS) in the mid infrared region has been applied to investigate chemical reaction kinetics in dc plasmas as well as in planar microwave discharges. It has been proofed to be a versatile and powerful diagnostic method well suited for such purposes.

It was possible to determine absolute ground state concentrations of nine stable molecular species (CH_4, CH_3OH, C_2H_2, C_2H_4, C_2H_6, NH_3, HCN, CH_2O, C_2N_2) and, in particular, of the methyl (CH_3) and of the cyanic radical (CN) in microwave discharges containing hydrogen, nitrogen and argon with small admixtures of methane or methanol under a wide range of gas mixing conditions. The measurements were combined with chemical modelling of the plasma, the ultimative objective being a better understanding of the chemical and reaction kinetic processes occurring in the plasma. The degree of dissociation of the precursor hydrocarbons varied between 20 and 97 %. It increased with the percentage of nitrogen in the flow. The concentration of the methyl radical was measured to be in the range of $10^{12} - 10^{13}$ molecules cm^{-3} while the cyanic radical was estimated to be between 10^{10} and 10^{12} molecules cm^{-3} depending on the discharge conditions. Measurements of the temporal development of molecular species concentrations while changing from flowing to static discharge conditions showed that hydro cyanic acid (HCN) and ammonia (NH_3) are the main final products of the chemical conversion processes in the discharge. Methane was another important conversion product in the case of methanol as precursor hydrocarbon. Therefore, in addition to the precursor hydrocarbons the highest concentrations ($10^{12} - 10^{15}$ molecules cm^{-3}) are measured for these three species followed by the stable and unstable hydrocarbon products (C_2H_2, C_2H_4, C_2H_6 and CH_3) with concentration in the range $10^{11} - 10^{14}$ molecules cm^{-3}. More common properties of discharges containing methane or methanol were found to be the following:

(a)   The degree of dissociation of the hydrocarbon precursor is higher for static discharge conditions (70…95 %) than with flowing gas (20…70 %).

(b)   The dominant final product species are HCN and NH_3. They contribute at most a few percent to the total pressure in the discharge.

(c)   Under all conditions, C_2H_4 showed lower concentrations than C_2H_2 or C_2H_6.

(d)   The concentration of the C2-products decreases with the nitrogen admixtures while the main final products HCN and NH$_3$ have their maximum concentrations at medium hydrogen to nitrogen ratios.

Some species could only be measured for one of the hydrocarbon precursor admixtures. The most significant differences in the plasma chemical reactions between methane or methanol as hydrocarbon precursors in H$_2$-Ar-N$_2$ microwave discharges are:

(I)   The degree of dissociation is higher for methanol than for methane in the gas mixture.

(II)   Methane is a dominant product of methanol containing discharges with concentrations of up to 10^{14} molecules cm^{-3} in the absence of nitrogen.

(III)   While in the methane case the concentrations of C$_2$H$_2$, CH$_3$ and C$_2$H$_6$ are comparable for all H$_2$ / N$_2$-ratios, in the methanol case the concentrations of CH$_3$ and C$_2$H$_6$ are several times higher for higher nitrogen contents.

(IV)   Although under flowing conditions higher CH$_3$ concentrations were observed with methanol as precursor, for static discharge conditions more CH$_3$ was measured in the methane case.

(V)   H$_2$CO was only observed for methanol admixtures, with concentrations in the same range as the product hydrocarbons of $10^{11} - 10^{14}$ molecules cm^{-3}.

(VI)   CN and C$_2$N$_2$ are formed with methane in gas flow with rather low concentration values of $10^{10} - 10^{12}$ molecules cm^{-3}.

The mass balance of carbon was determined for both methane-containing and methanol-containing plasmas. Up to 30 % of the available carbon was bound in HCN. Only for small amounts or no nitrogen in the gas mixture hydrocarbons, in particular C$_2$H$_2$ and C$_2$H$_6$ contribute in maximum with several percents significantly to the mass balance. In the methanol case the methane produced contributes additionally in the order of some percents.

From these results several conclusions can be made for such discharges of mixtures of hydrogen, nitrogen and argon with small additions of methane or methanol:

(I)   The degree of dissociation increases while changing to higher nitrogen contents and with methanol instead of methane in the gas mixture. This effect is caused most likely by the higher number of different active species produced under these circumstances.

(II)   Not ammonia, as one would expect, is the most dominant product species but hydro cyanic acid. This proofs that the admixture has a significant influence on the chemistry in these discharges.

(III) With methanol a higher degree of dissociation of the precursor hydrocarbon, but lower concentration of the two main products was observed under similar discharge conditions. This together with the higher amount of produced hydrocarbon species can indicate that for methanol a more reactive and more species rich plasma is established.

Some other important intermediates, in particular NH, NH_2 or CH_2, should be observable using the present absorption experiment i.e. with a multiple pass arrangement providing an interaction length in the order of several meters in the active plasma zone. The extension of TDLAS measurements to these species is in progress, but since their line strengths are presently unknown, the determination of their absolute concentrations is impossible for the moment.

Based on the measured degrees of dissociation and concentration values fragmentation rates of the two hydrocarbon precursors and conversion rates to the main products of the reaction kinetics were estimated. The fragmentation rate of methane increases with the nitrogen content from 3×10^{15} molecules J^{-1} to 7×10^{15} molecules J^{-1}. The fragmentation rate of methanol was slightly higher. It ranged from 6×10^{15} molecules J^{-1} to 9×10^{15} molecules J^{-1} increasing with the nitrogen percentage in the flow as well. The rates of conversion to the main final products HCN and NH_3 as well as the conversion rates to hydrocarbon products are slightly smaller. They range between $10^{13} - 10^{15}$ molecules J^{-1}, while HCN has the largest conversion rate.

An extended model of the chemical reaction kinetics has been established for H_2-N_2-Ar-CH_4 plasmas. The model calculations were performed including 22 species and using a set of 145 chemical reactions and appropriate electron impact dissociation rate coefficients. The results of the model calculations showed a satisfying agreement with measured concentrations within the order of errors of measurements. On the basis of these calculation a simplified reaction scheme of the H_2-N_2-Ar-CH_4 reaction kinetics was proposed describing the complex chemistry in a good way. The results of the model calculations were used to predict concentrations of non-measured species like NH, NH_2 or CH_2, which were expected to have concentrations in the order of 10^{11} to 10^{12} molecules cm^{-3}. So, in further investigations, combining absorption spectroscopic methods with multiple pass optics, it should be possible to investigate these radicals as well.

TDLAS was used for spectroscopic investigations of the fundamental bands of $^{12}C^{14}N$ and $^{13}C^{14}N$ too. In preliminary diagnostic studies it was found, that the CN concentration depends strongly on the discharge power, pressure and nitrogen content in the gas mixture. Optimum discharge conditions for the production of CN were found to be at high powers (3 kW), low pressures (0.3 mbar) and equal nitrogen and methane content in the gas mixture (1 : 1). To get more insight into the reaction kinetics HCN was investigated for these discharge conditions as well. While both species showed a similar behaviour with power, an opposite behaviour was observed as a function of pressure.

18 ($^{12}C^{14}N$) and 22 ($^{13}C^{14}N$) P- and R-branch lines were detected and line positions determined with accuracy as high as 4×10^{-4} cm^{-1}. The measured line positions were used in five parameter fits giving molecular constants with an improved accuracy compared to earlier data. The origins of the fundamental bands of $^{12}C^{14}N$ (2042.42104(84) cm^{-1}) and of $^{13}C^{14}N$ (2000.08470(30) cm^{-1}) agree with v_0 values in the literature within their 3σ values. A fine structure was observed only for some lines. This is caused by an experimental Doppler line width in the order of the fine splitting. Fine splitting therefore was ignored for the fitting if observed and average line position used instead.

Absorption spectroscopic methods were applied to dc discharges as well. In particular the conversion kinetics of CO_2 to CO in the stationary afterglow of closed dc glow discharge in CO_2 were observed. For this purpose, absolute ground state concentrations of CO_2 as well as of CO were measured in the stationary afterglow time and spatially resolved for 300 s until a so-called quasi-equilibrium state of stable reaction product densities was established. Finally between 40 to 60 % of CO_2 was converted to CO more predominantly with increasing discharge current. The formation time constant of the stable gas composition decreased with the discharge current as well. The spatial dependence of the equilibrium CO concentration was found to be varying less than 10 %. For this chemical system a conversion rate was estimated as well. It was found to be about 1.2×10^{13} molecules J^{-1} for currents higher than 10 mA. Based on the experimental results a model of the plasma chemical reaction kinetics was derived describing the temporal development of both species with satisfying accuracy.

Besides, TDLAS was used to investigate the conversion of $TiCl_4$ to HCl in a pulsed H_2-Ar-N_2-$TiCl_4$ dc-reactor. The gas well was of 417 sccm H_2, 89 sccm Ar and 77 sccm N_2 with $TiCl_4$ flows of 1.8 sccm in maximum. The total pressure was fixed to 2 mbar. It was proofed, that for wide range of $TiCl_4$ admixtures up to 1 sccm (>0.17 % in the total gas flux) Cl was completely transferred into HCl in the gas phase. For higher admixtures a saturation effect was observed. The HCl concentration was here about 4...6 x 10^{14} molecules cm^{-3}, significantly depending on the discharge voltage. In contradiction to that, for smaller $TiCl_4$ no dependence of the HCl concentrations on the discharge voltage was found. Therefore $TiCl_4$ amounts of 1 sccm and average discharge voltages should be preferred for optimised discharge conditions in the future.

9   References

[1]   Haverlag M, Stoffels E, Stoffels W W, Kroesen G M W and De Hoog F J 1996 *J. Vac. Sci. Technol.* A **14** 380

[2]   Röpcke J, Mechold L, Käning M, Fan W Y and Davies P B 1999 *Plasma Chem. Plasma Process.* **19** 395

[3]   Davies P B and Martineau P M 1992 *Adv. Mater.* **4** 729

[4]   Naito S, Ito N, Hattori T and Goto T 1995 *Jpn. J. Appl. Phys.* **34,** 302

[5]   Haverlag M, Stoffels E, Stoffels W W, Kroesen G M W and De Hoog F J 1994 *J. Vac. Sci. Technol.* A **12** 3102

[6]   Kawaguchi K, Endo Y and Hirota E 1982 *J. Mol. Spectr.* **93** 381

[7]   Hunt N T, Röpcke J and Davies P B 2000 *J. Mol. Spectr.* **204** 120

[8]   Liu Z and Davies P B 1996 *J. Chem. Phys.* **105** 3443

[9]   Osiac M, Röpcke J and Davies P B 2001 *Chem. Phys. Lett.* **344** 92

[10]   Fan W Y, Knewstubb P F, Käning M, Mechold L, Röpcke J and Davies P B 1999 *J. Phys. Chem.* A **103** 4118

[11]   Mechold L, Röpcke J, Duten X and Rousseau A 2001 *Plasma Sources Sci. Technol.* **10** 52

[12]   Röpcke J, Mechold L, Duten X and Rousseau A 2001 *J Phys. D: Appl. Phys.* **34** 2336

[13]   Busch C, Möller I and Soltwisch H 2001 *Plasma Sources Sci. Technol.* **10** 250

[14]   Röpcke J, Revalde G, Osiac M, Li K and Meichsner J 2002 *Plasma Chem. Plasma Process.* **22** 137

[15]   Vandevelde T, Nesladek M, Quaeyhaegens C and Stals L 1997 *Thin Solid Films* **308-309** 154

[16]   Vandevelde T, Wu T D, Quaeyhaegens C, Vlekken J, D'Olieslaeger M and Stals L 1999 *Thin Solid Films* **340** 159

[17]   Mutsukura N 2001 *Plasma Chem. Plasma Process.* **21** 265

[18]   Bhattacharyya S, Granier A and Turban G 1999 *J. Appl. Phys.* **86** 4668

[19] Zhang M, Nakayama Y, Miyazaki T and Kume M 1998 *J. Appl. Phys.* **85** 2904

[20] Dinescu G, De Graaf A, Aldea E, and van de Sanden M C M 2001 *Plasma Sources Sci. Technol.* **10** 513

[21] De Graaf A, Aldea E, Dinescu G and Van de Sanden M C M 2001 *Plasma Sources Sci. Technol.* **10** 524

[22] Penetrante B M, Hsiao M C; Bardsley J N; Merritt B T, Vogtlin G E, Kuthi A, Burkhart C P and Bayless J R 1997 *Plasma Sources Sci. Technol.* **6** 251

[23] Kareev M, Sablier M and Fujii T 2000 *J. Phys. Chem.* **104** 7218

[24] Coll P, Coscia D, Gazeau M C, De Vanssay E, Guillemin J C and Raulin F 1995 *Adv. Space Res.* **16** 93

[25] De Vanssay E, Gazeau M C, Guillemin J C and Raulin F 1995 *Planet. Space Sci.* **43** 25

[26] Gupta S, Ochiai E and Ponnamperuma C 1981 *Nature* **293** 725

[27] Raulin F, Mourey D and Toupance G 1982 *Origins of Life* **12** 267

[28] Thompson W R, Henry T J, Schwartz J M, Khare B N and Sagan C 1991 *Icarus* **90** 57

[29] Cabane M and Chassefiere E 1995 *Planet. Space Sci.* **43** (1/2) 47

[30] Toupance G, Raulin F and Bouvet R 1975 *Origins of Life* **6** 83

[31] Tabares F L, Tafalla D, Tanarro I, Herrero V J, Islyaikin A and Maffiotte C 2002 *Plasma Phys. Control. Fusion* **44** L1

[32] Zarrabian M, Leteinturier C and Turban G 1998 *Plasma Sources Sci. Technol.* **7** 607

[33] Sugai H, Kojima H, Ishida A and Toyoda H 1990 *Appl. Phys. Lett.* **56** 2616

[34] Sugai H and Toyoda H 1992 *J. Vac. Sci Technol.* A **8** 1193

[35] Schmidt M, Foest R and Basner R 1998 *J. Phys. IV France* **8** 231

[36] Celii F G and Butler J E 1992 *J. Appl. Phys.* **71** 2877

[37] Ando S, Shinohara M and Takayama K 1998 *Vacuum* **49** 113

[38] Childs M A, Menningen K L, Chevako P, Spellmeyer N W, Anderson L W and Lawler J E 1992 *Phys. Lett.* A **171** 87

[39] Zalicki P, Ma Y, Zare R N, Wahl E H, Dadamio J R, Owano T G, Kruger C H 1995 *Chem. Phys. Lett.* **234** 269

[40] Tanarro I, Sanz M M, Bermejo D, Domingo C and Santos J 1994 *J. Chem. Phys.* **100** 238

[41] Yamada C, Hirota E and Kawaguchi K 1981 *J. Chem. Phys.* **75** 5256

[42] Yamada C and Hirota E 1983 *J. Chem. Phys.* **78** 669

[43] Nelson D D *et al* 1996 *SPIE Proceedings* **2834** 148

[44] Jimenez J *et al* 1999 *J. Air & Waste Manage. Asso.* **49** 463

[45] Schiff H I, Chanda A, Pisano J and Mackay G 1998 *SPIE Proceedings* **3535** 132

[46] Röpcke J, Mechold L, Käning M, Anders J, Wienhold F G, Nelson D and Zahniser M 2000 *Rev. Scient. Instr.* **71** 3706

[47] McManus J B, Mechold L, Nelson D, Osiac M, Röpcke J, Rousseau A and Zahniser M 2002 ESCAMPIG 16 / ICRP 5, Grenoble, N. Sadeghi, H. Sugai Eds., **2**. 279

[48] Ohl A, *Large Area Planar Microwave Discharges*, in *Microwave Discharges: Fundamentals and Applications*, C M Ferreira and M. Moisan (Ed.), Plenum, New York, 1993, 205

[49] Ohl A and Röpcke J 1992 *Diam. Relat. Mater.* **I** 243

[50] Ohl A, Röpcke J, and Schleinitz W 1993 *Diam. Relat. Mater.* **2** 298

[51] Röpcke J, Ohl A and Schmidt M 1993 *J. Analyt. Atomic Spectrom.* **8** 803

[52] Ohl A, Strobel H, Röpcke J, Kammerstetter H, Pries A and Schneider M 1995 *Surf. Coat. Technol.* **74-75** 59

[53] Keller D, Schröder K, Husen B and Ohl, A 1997 *Polymer Preprints* **38** 1043

[54] Lavrov B P, Melnikov A S, Käning M and Röpcke J 1999 *Phys. Rev. E* **59** 3526

[55] Osiac M, Lavrov B P and Röpcke J 2002 *J. Quant. Spectrosc. Radiat. Transf.* **74** 471

[56] Mechold L, *Diagnostic Studies of Microwave Plasmas Containing Hydrocarbons Using Tunable Diode Lasers*, Dissertation, Ernst-Moritz-Arndt Universität Greifswald 2000

[57] Hippler R, Pfau S, Schmidt M and Schoenbach K H, *Low Temperature Plasma Physics*, Wiley-Vch Verlag Berlin, Germany, 2001

[58] Langmuir I and Mott-Smith H M 1923 *Gen. Elec. Rev.* **26** 731

[59]   Allen J E, Boyd R L F and Reynolds P 1957 *Proc. Phys. Fluids* **2** 112

[60]   Bohm D, *The characteristic of electrical discharges in magnetic field* ch.3, Ed. Guthrie and Mc Graw Hill Book Co. Inc., 1949

[61]   Laframboise J G, Univ. Toronto. UTIAS Rept. No. 100, 1966

[62]   Chen F F, *Electric Probes* in *Plasma Diagnostic Techniques*, Ed. Huddlestone R H and Leonard S L, Academic Press, New York, 1965

[63]   Zakrzewski Z and Kopiczynski T 1974 *Plasma Phys.* **16** 1194

[64]   Rousseau A, Teboul E, Lang N, Hannemann M and Röpcke J 2002 *J. Appl. Phys.* **92** 3463

[65]   Schmidt M, Foest R and Basner R, *Mass spectrometric diagnostics* in *Low Temperature Plasma Physics*, edited by Hippler R, Pfau S, Schmidt M and Schoenbach K H, Wiley-Vch Verlag Berlin, Germany, 2001, 199

[66]   Darwin H W, *Mass Spectrometry of Plasmas* in *Plasma Diagnostics*, edited by Lochte-Holtgreven W, Amsterdam, 1968, 777

[67]   Schmidt M and Hinzpeter G 1970 *Beitr. Plasmaphys.* **10** 183

[68]   Vasile M J and Dylla H F, *Mass Spectrometry in Plasmas* in *Plasma Diagnostics*, edited by Auciello O and Flamm D L, Academic Press Boston, Vol. 1, 1989, 185

[69]   Paul W and Steinwedel H 1953 *Z. Naturforschg.* **8a** 448

[70]   Basner R, Foest R, Schmidt M, Sigeneger F, Kurunczi P, Becker K and Deutsch H 1996 *Int. J. of Mass Spectr. And Ion Process.* **153** 65

[71]   Jauberteau J L, Thomas L, Aubreton J, Jauberteau I and Catherinot A 1998 *Plasma Chem. Plasma Process.* **18** 137

[72]   Benndorf C, Joeris P and Kröger R 1994 *Pure & Appl. Chem.* **66** 1195

[73]   Berkowitz J, *Photoabsorption, Photoionization and Photoelectron Spectroscopy*, Academic Press, New York, 1979

[74]   Rödel W and Wölm G, *Grundlagen der Gaschromatographie*, DVW, Berlin, 1982

[75]   Röpcke J, Davies P B, Käning M and Lavrov B P, *Spectroscopical Diagnostics of Molecular Non-Equilibrium Plasmas* in *Low Temperature Plasma Physics*, edited by

Hippler R, Pfau S, Schmidt M and Schoenbach K H, Wiley-Vch Verlag Berlin, Germany, 2001, 173

[76] Coburn J W and Chen M 1980 *J. Appl. Phys.* **51** 3134

[77] Coburn J W and Chen M 1981 *J. Vac. Sci. Technol.* **18** 353

[78] Röpcke J and Ohl A 1991 *Contrib. Plasma Phys.* **31** 669

[79] Röpcke J and Ohl A 1994 *Contrib. Plasma Phys.* **34** 575

[80] Demtröder W, *Laserspektroskopie*, Springer-Verlag, Berlin, 1991

[81] De Avillez Pereira R, Baulch D L, Pilling M J, Robertson S H and Zeng G 1997 *J. Phys. Chem. A* **101** 9681

[82] Berman M R, Fleming J W, Harvey A B and Lin M C 1982 *Chem Phys.* **73** 27

[83] Berman M R and Lin M C 1984 *J. Chem. Phys.* **81** 5743

[84] Hummernbrum F, Kempkens H, Ruzicka A, Sauren H D, Schiffer C, Uhlenbusch J and Winter J 1992 *Plasma Sources Sci. Technol.* **1** 221

[85] Doerk T, Hertl M, Pfelzer P, Hädrich S, Jauernik P, Uhlenbusch J 1997 *Appl. Phys.* **B64** 111

[86] Lichtin D A and Lin M C 1985 *Chem. Phys.* **96** 473

[87] Lichtin D A and Lin M C 1986 *Chem. Phys.* **104** 325

[88] Czarnetzki U, Miyazaki K, Kajiwara T, Muraoka K, Maeda M and Döbele H F 1994 *Appl. Opt.* **11** 2155

[89] Duan X R, Lang N, Lange H and Röpcke J, *9th Intern. Symposium on Laser-Aided Plasma Diagnostics, Lake Tahoe, Conf. Proc.,* 1999, 217

[90] Duan X R and Lange H, *Verhandlungen der DPG, Frühjahrstagung*, Bonn, 2000, 1023

[91] Druet S A J and Taran J P E 1981 *Prog. Quantum Electron.* **7** 1

[92] Eckbreth A C and Hall R J 1981 *Combustion Sci. Technol.* **25** 175

[93] Eckbreth A C, *Laser Diagnostics for Combustion Chemistry*, 2nd Edition, Ch. 6, Gordon and Breach Publishers, Amsterdam, 1996

[94] Kempkens H and Uhlenbusch J 2000 *Plasma Sources Sci. Technol.* **9** 492

[95] Doerk T, Ehlbeck J, Jedamzik R, Uhlenbusch J, Höschele J and Steinwandel J 1997
      Appl. Spectrosc. **51** 1360

[96] Pott A, Doerk T, Uhlenbusch J, Ehlbeck J, Höschele J and Steinwandel J 1998 *J. Phys.*
      D: Appl. Phys. **31** 2485

[97] Baeva M, Luo X, Pfelzer B, Schäfer J H, Uhlenbusch J and Zhang Z 1999 *Plasma*
      Sources Sci. Technol. **8** 142

[98] Baeva M, Gier H, Pott A, Uhlenbusch J, Höschele J and Steinwandel J 2001 *Plasma*
      Chem. Plasma Proc. **21** 225

[99] Baeva M, Gier H, Pott A, Uhlenbusch J, Höschele J and Steinwandel J 2002 *Plasma*
      Sources Sci Technol. **11** 1

[100] Baeca M, Ehlbeck J, Pott A, Repsilber T and Uhlenbusch J 2002 *SPIE Proceedings*
      4460 134

[101] Hädrich S, Pfelzer B and Uhlenbusch J 1999 *Plasma Chem. Plasma Process.* **19** 91

[102] Kornas V, Schulz-von der Gathen V, Bornemann T and Döbele H F 1991 *Plasma*
      Chem. Plasma Process. **11** 171

[103] Kornas V, Roth A, Döbele H F and Pross G 1995 *Plasma Chem. Plasma Process.* **15**
      71

[104] Conrads J P F and Bandlow I, *Mathematisch-numerische und experimentelle*
      Untersuchungen zur Absorptionstomographie mit Multipathanordnung als
      diagnostische Methode für Niederdruckplasmen, DFG-Abschlußbericht, CO 233/2-1,
      2001

[105] O'Keefe A and Deacon D A G 1988 *Rev. Sci. Instr.* **59** 2544

[106] Orr-Ewing A J, *Pulsed and continuous wave cavity ring-down spectroscopy probes of*
      chemical vapour deposition plasmas, Frontiers in Low Temperature Plasma
      Diagnostics IV, Rolduc 2001, Conf. Proc., pp. 30-39

[107] Romanini D and Lehmann K K 1993 *J. Chem. Phys.* **99** 6287

[108] Booth J P, Cunge G, Biennier L, Romanini D and Kachanov A 2000 *Chem. Phys. Lett.*
      317 631

[109] Engeln R, Letourneur K G Y, Boogaarts M G H, van de Sanden M C M and Schram D
      C 1999 *Chem. Phys. Lett.* **310** 405

[110] Zalicki P, Ma Y, Zare R N, Wahl E H, Owano T G and Kruger C H 1995 *Appl. Phys. Lett.* **67** 144

[111] Hemerik M M and Kroesen G M W, *The final frontiers of cavity ring down spectroscopy: the mid-infrared*, Frontiers in Low Temperature Plasma Diagnostics IV, Rolduc 2001, Conf. Proc. pp. 83-86

[112] Hanst P L, *Gas Analysis – Topics and Products*, Infrared Analysis Booklet and Catalog 2000

[113] Grisar R, *Quantitative Gasanalyse mit abstimmbaren IR-Diodenlasern*, IPM-Forschungsbericht 24-4-92

[114] Guelachvili G and Rao K N, *Handbook of Infrared Standards*, Academic Press, Orlando, 1986

[115] Maki A G and Wells J S, *Wavenumber Calibration Tables From Heterodyne Frequency Measurements*, NIST Special Publication 821, 1991

[116] Rothman L S, Gamache R R, Tipping R H, Rinsland C P, Smith M A H, Benner D C, Malathy Devi V, Flaud J-M, Camy-Peyret C, Perrin A, Goldman A, Massie S T, Brown L R and Toth R A, 1992 *J. Quant. Spectrosc. & Radiat. Transfer* **48** 469

[117] Rothman L S, Rinsland C P, Goldman A, Massie S T, Edwards D P, Flaud J-M, Perrin A, Camy-Peyret C, Dana V, Mandin J-Y, Schroeder J, McCann A, Gamache R R, Wattson R B, Yoshino K, Chance K V, Jucks K W, Brown L R, Nemtchinov V and Varanasi P 1998 *J. Quant. Spectrosc. Radiat. Transfer* **60** 665

[118] Husson N, Bonnet B, Chedin A, Scott N A, Chursin A A, Golovko V F and Tyuterev V G 1994 *J. Quant. Spectrosc. & Radiat. Transfer* **52** 425

[119] Jacquinet-Husson N, Scott N A, Chedin A, Bonnet B, Barbe A, Tyuterev V G, Champion J P, Winnewisser M, Brown L R, Gamache R, Golovko V F and Chursin A A 1998 *J. Quant. Spectrosc. Radiat. Transfer* **59** 511

[120] Jacquinet-Husson N, Ariés E, Ballard J, Barbe A, Bjoraker G, Bonnet B, Brown L R, Camy-Peyret C, Champion J P, Chédin A, Chursin A, Clerbaux C, Duxbury G, Flaud J-M, Fourrié N, Fayt A, Graner G, Gamache R, Goldman A, Golovko V, Guelachvili G, Hartmann J M, Hilico J C, Hillman J, Lefàvre G, Lellouch E, Mikhailenko S N, Naumenko O V, Nemtchinov V, Newnham D A, Nikitin A, Orphal J, Perrin A, Reuter D C, Rinsland C P, Rosenmann L, Rothman L S, Scott N A, Selby J, Sinitsa L N, Sirota

J M, Smith A M, Smith K M, Tyuterev V G, Tipping R H, Urban S, Varanasi P and Weber M 1999 *J. Quant. Spectrosc. & Radiat. Transfer* **62** 205

[121] Drost H, *Plasmachemie*, Akademie Verlag, Berlin, 1978

[122] *Plasma Technology – Process Diversity + Sustainability*, German Federal Ministry of Education and Research, Department of Public Relations, 2001

[123] Conrads H and Schmidt M 2000 *Plasma Sources Sci. Technol.* **9** 441

[124] Moisan M and Pelletier J, *Microwave excited Plasma*, *Plasma Technology* **4**, Elsevier Science Publishers B.V., Netherlands, 1992

[125] Wertheimer M R and Moisan M 1985 *J. Vac. Sci. Technol. A* **3** 2643

[126] Ferreira C M and Moisan M, *Microwave excited Plasma*, *Plasma Technology* **4**, Chapter 3, Elsevier Science Publishers B.V., Netherlands, 1992

[127] Moisan M, Barbeau C, Claude R, Ferreira C M, Margot J, Paraszczak J, Sa A B, Sauve G and Wertheimer M R 1991 *J.Vac. Sci. Technol. B* **9** 8

[128] Winkler R, *Collision Dominated Electron Kinetics in Low and High Frequency Fields* in *Microwave Discharges: Fundamentals and Applications*, Edited by Ferreira C M and Moisan M, 1991, 339

[129] Marec J and Leprince P 1998 *J. Phys. IV France* **8** Pr7-1

[130] Rint C, Herausgeber, *Handbuch für Hochfrequenz und Elektro-Techniker*, Band 2, Kapitel 2, Hütig und Pflaum Verlag, 1978, 612

[131] Kummer M, *Grundlagen der Mikrowellentechnik*, Kapitel *Wellenleiter*, VEB Verlag Technik, 1986, 76

[132] Käning M, *Emissionsspektroskopische Charakterisierung von Mikrowellenplasmen*, Dissertation, Ernst-Moritz-Arndt-Universität Greifswald, 1997

[133] Ohl A 1998 *J. Phys. IV France* **8** Pr7-82

[134] Ohl A 1994 *Pure & Appl. Chem.* **66** 1397

[135] White J U 1942 *J. Optical Soc. America* **32** 285

[136] Röpcke J, Käning M and Lavrov B P 1998 *J. Phys. IV* **8** 207

[137] Astashkevich S A, Käning M, Käning E, Kokina N V, Lavrov B P, Ohl A and Röpcke J 1996 *J. Quant. Spectrosc. Radiat. Transfer* **56** 725

[138] Röpcke J, Lavrov B P and Ohl A, *Spectroscopical investigations of planar microwave discharges*, Frontiers in Low Temperature Plasma Diagnostics II, Bad Honnef, 1997, Book of Papers, 201

[139] Lavrov B P, Käning M, Ovtchinnikov V L and Röpcke J, *Fine structure of Balmer-lines and determination of translational temperature of hydrogen and deuterium containing plasmas*, Frontiers in Low Temperature Plasma Diagnostics II, Bad Honnef, 1997, Book of Papers, 169

[140] Gemini Scientific Instr. 2001 *Multi-Pass Optics Alignment Instructions*, Manual

[141] Zahniser M S, Nelson D D and Kolb C E 2002 in *Applied Combustion Diagnostics*, Kohse-Hoinghaus K and Jeffries J (eds), Tallor and Francis, New York, 648

[142] Nelson D D, Shorter J H, McManus J B and Zahniser M S 2002 Appl. Phys. B in print.

[143] Wormhoudt J and McCurdy K E 1989 *Chem. Phys. Lett.* **156** 47

[144] Balla R J and Pasternak L 1987 *J. Phys. Chem.* **91** 73

[145] Herzberg G and Shoosmith J 1956 *Can. J. Phys.* **34** 523

[146] Herzberg G 1961 *Proc. Roy. Soc. A* **262** 291

[147] Amano T, Bernath P F, Yamada C, Endo Y and E. Hirota E 1982 *J. Chem Phys.* **77** 5284

[148] Bethardy G A, Northrup F J and Macdonald R G 1996 *J. Chem. Phys.* **105** 5433

[149] Fairbrother D H, Briggman K A, Dickens K A, Stair P C and Weitz E 1997 *Rev. Sci. Instrum.* **68** 2031

[150] Robinson G N, Zahniser M S, Freedman A and Nelson Jr. D D 1996 *J. Mol. Spectr.* **176** 337

[151] Takayanagi T 1996 *J. Chem. Phys.* **104** 2237

[152] Tachibana K, Nishida M, Harima H and Urano Y 1984 *J. Phys. D: Appl. Phys.* **17** 1727

[153] Celii F G, Pehrsson P E, Wang H T and Butler J E 1988 *Appl. Phys. Lett.* **52** 2043

[154] Wormhoudt J 1990 *J. Vac. Sci. Technol. A* **8** 1722

[155] Ikeda M, Aiso K, Hori M and Goto T 1995 *Jpn. J. Appl. Phys.* **34** 3273

[156] Ikeda M., Hori M, Goto T, Inayoshi M, Yamada K, Hiramatsu M and Nawata M 1995 *Jpn. J. Appl. Phys.* **34** 2484

[157] Naito S, Ito N, Hattori T and Goto T 1994 *Jpn. J. Appl. Phys.* **33** 5967

[158] Naito S, Ikeda M, Ito N, Hattori T and Goto T 1993 *Jpn. J. Appl. Phys.* **32** 5721

[159] Ikeda M, Ito N, Hiramatsu M, Hori M and Goto T 1997 *J. Appl. Phys.* **82** 4055

[160] Lombardi G, Stancu G D, Hempel F, Gicquel A and Röpcke J 2003 *to be published*

[161] Bohr S, Haubner R and Lux B 1996 *Appl. Phys. Lett.* **68** 1075

[162] Dagel D J, Mallouris C M and Doyle J R 1996 *J. Appl. Phys.* **79** 8735

[163] Gardner W L 1996 *J. Vac. Sci. Technol. A* **14** 1938

[164] Harris S and Weiner A M 1990 *J. Appl. Phys.* **67** 6520

[165] Pauser H, Schwärzler C G, Laimer J and Störi H 1997 *Plasma Chem. Plasma Process.* **17** 107

[166] Schwärzler C G, Schnabl O, Laimer J and Störi H 1996 *Plasma Chem. Plasma Process.* **16** 173

[167] Kline L E, Partlow W D and Bies W E 1989 *J. Appl. Phys.* **65** 70

[168] Hempel F, Davies P B, Loffhagen D, Mechold L and Röpcke J 2003 *to be published*

[169] Buckman S J and Phelps A V 1985 *J. Chem. Phys.* **82** 4999

[170] Phelps A V and Pitchford L C 1985 *Phys. Rev.* **31** 2932

[171] Ehrhardt A B and Langer W D, *Collisional Processes of Hydrocarbons in Hdrogen Plasmas*, Princeton Plasma Phys. Laboratory Report – PPPL-2477, 1987

[172] Brooks J N, Wang Z, Ruzic D N and Alman D A, *Hydrocarbon Rate Coefficients for Proton and Electron Impact Ionisation, Dissociation, and Recombination in a Hydrogen Plasma*, PANF/FPP/TM-297, 1999

[173] Baulch D L, Cobos C J, Cox R A, Esser C, Frank P, Just T, Kerr J A, Pilling M J, Troe J, Walker R W and Warnatz, J 1992 *J. Phys. Chem. Ref. Data* **21** 411

[174] Warnatz J, *Rate coefficients in the C/H/O system* in *Combustion Chemistry*, ed. Gardiner Jr. W C, Springer-Verlag, NY 1984, 197

[175] Fan W Y, *Plasma Diagnostics and Spectroscopy Using Tunable IR Diode Lasers*, Dissertation Thesis, University of Cambridge, Department of Chemistry, 1997

[176] Legrand J C, Diamy A M, Hrach R and Hrachova V 1997 *Contrib. Plasma Phys.* 1997 **6** 521

[177] Tsang W and Hampson R F 1986 *J. Phys. Chem. Ref. Data* **15** 1087

[178] Gordiets B, Ferreira C M, Pinheiro M J and Ricard A 1998 *Plasma Sources Sci. Technol.* **7** 363

[179] Herron J T and Green D S 2001 *Plasma Chem. Plasma Process.* **21** 459

[180] Hack W, *NH Radical Reactions* in *N-Centered Radicals,* ed. Alfassi Z B, John Wiley & Sons, Chichester, 413

[181] McDaniel, A H and Allendorf M D 1998 *J. Phys. Chem. A*

[182] Smith G P et al, *Gri-Mech,* http://www.me.berkeley.edu/gri_mech/

[183] Miller J A and Bowman C T 1989 *Prog. Energy Combust. Sci.* **15** 287

[184] Deppe J, Friedrichs G, Römming H J and Wagner H G 1999 *Phys. Chem. Chem. Phys.* **1** 427

[185] Baulch D L, Duxbury J, Grant S J and Montague D C 1981 *J. Phys. Chem. Ref. Data* **10** 1

[186] Tahara H, Minami K I, Murai A, Yasui T and Yoshikawa T 1995 *Jpn. J. Appl. Phys* **34** 1972

[187] Oumghar A, Legrand J C, Diamy A M and Turillon N 1995 *Plasma Chem. Plasma Process.* **15** 87

[188] Baulch D L, Cobos C J, Cox R A, Frank P, Hayman G, Just T, Kerr J A, Murrells T, Pilling M J, Troe J, Walker R W and Warnatz J 1994 *J. Phys. Chem. Ref. Data* **23** 847

[189] Dean A J, Hanson R K and Bowman C T 1991 *J. Phys. Chem.* **95** 3180

[190] Melnikov A, *Investigations of the concentration of hydrogen atoms in plasma of planar microwave discharge by laser induced fluorescence method*, INP-Report, 2001

[191] Rehfuss B D, Suh M H, Miller T A and Bondybey V E 1992 *J. Mol. Spec.* **151** 437

[192] Prasad C V V, Bernath P F, Frum C and Engleman R 1992 *J. Mol. Spec.* **151** 459

[193] Skatrud D D, DeLucia F C, Blake G A and Sastry K V L N 1983 *J. Mol. Spec.* **99** 35

[194] Bogey M, Demuynck C and Destombes J L 1986 *Chem. Phys.* **102** 141

[195] Penzias A A, Wilson R W and Jefferts K B 1974 *Phys. Rev. Lett.* **32** 701

[196] Lambert D L, Brown J A, Hinkle K H and Johnson H R 1984 *Astrophys. J.* **284** 223

[197] Feher M, Rohrbacher A and Maier J P 1993 *Chem. Phys.* **173** 187

[198] Treffers R R 1975 *Astrophys. J.* **196** 883

[199] Davies P B and Hamilton P A 1982 *J. Chem. Phys.* **76** 2127

[200] Davis S P, Abrams M C, Rao M L P and Brault J W 1991 *J. Opt. Soc. Am. B* **8** 198

[201] Pichamuthu J P, *Coherent Sources and Applications* in *IR and mm Waves* Vol. 7, Part II, ed. Button K J, Academic, NY, 1983, 165

[202] Skatrud D D and DeLucia F C 1985 *Appl. Phys. Lett.* **46** 631

[203] Cerny D, Bacis R, Guelachvili G and Roux F 1978 *J. Mol. Spec.* **73** 154

[204] Hosinsky G, Klynning L and Lindgren B 1982 *Phys. Scrip.* **25** 291

[205] Poole C P in *Electron Spin Resonance*, John Wiley and Sons, New York, 2nd Edition, 1983, Ch. 12

[206] Smith D M and Davies P B 1994 *J. Chem. Phys.* **100** 6166

[207] Linnerud I, Kaspersen P, Jæger T 1998 *Appl. Phys. B* **67** 297

[208] Mihalcea R M, Baer D S and Hanson R K 1997 *Appl. Optics* **36** 8745

[209] Mihalcea R M, Baer D S and Hanson R K 1998 *Meas. Sci. Technol.* **9** 327

[210] Mihalcea R M, Webber M E, Baer D S, Hanson R K, Feller G S, Chapman W B 1998 *Appl. Phys. B* **67** 283

[211] Daniel R G, McNesby K L and Miziolek A W 1996 *Appl. Optics* **35** 4018

[212] Schoenung S M and Hanson R K 1982 *Appl. Optics* **21** 1767

[213] Morvova M 1998 *J. Phys. D : Appl. Phys.* **31** 1865

[214] Giacobbe F W and Schmerling D W 1983 *Plasma Chem. Plasma Process.* **3** 383

[215] Kylian O, Leys C and Hrachova V 2001 *Contrib. Plasma Phys.* **41** 407

[216] Rutscher A and Wagner H-E 1993 *Plasma Sources Sci. Technol.* **2** 279

[217] Lucke W, Miethke F, Pfau S, Rutscher A and Wagner H-E 1993 *Proc. ISPC-11, Loughborough (England)* Vol. 3 1356

[218] Sonnenfeld A, Strobel H and Wagner H-E 1998 *J. Non-Equilib. Thermodyn.* **23** 105

[219] Gundermann S, Loffhagen, D, Wagner H-E and Winkler R 2001 *Contrib. Plasma Phys.* **41** 45

[220] Mechold L, Röpcke J, Käning M, Loffhagen D and Davies P B 1999 *Lausanne Report LRP* **629/99** 155

[221] Graves D B 1994 *IEEE Trans. Plasma Sci.* **22** 31

[222] Loffhagen D and Winkler R 1998 *8th Int. Symp. Sci. Tech. Light Sources* Greifswald Proc. 70

[223] Loffhagen D and Winkler R 1996 *J. Phys. D: Appl. Phys.* **29** 618

[224] Conrads H et.al., *Stand und Perspektiven der Plasmatechnologie*, Studie im Auftrage des BMBFT, FKZ 13N6182, 1995

[225] Tan L Y, Winer A M and Pimentel G C 1972 *J. Chem. Phys.* **57** 4028

10 Appendix

10.1 Additional figures

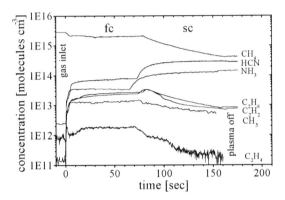

Figure 10.1: Temporal development of molecular concentrations in a methane containing discharge under flowing (fc) and static (sc) conditions for 405 sccm hydrogen gas flow rate

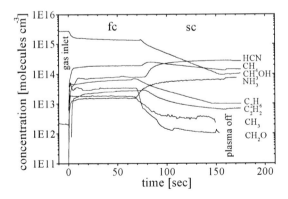

Figure 10.2: Temporal development of molecular concentrations in a methanol containing discharge under flowing (fc) and static (sc) conditions for 405 sccm hydrogen gas flow rate

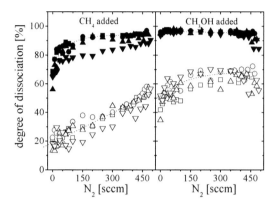

Figure 10.3: Degree of dissociation of methane and methanol under flowing (open symbols) and static (solid symbols) discharge conditions for hydrocarbon precursor admixtures of 40 sccm ($\square$,$\blacksquare$), 25 sccm ($\bigcirc$,$\bullet$), 15 sccm ($\triangle$,$\blacktriangle$) and 5 sccm ($\triangledown$,$\blacktriangledown$) and various H_2 / N_2-ratios

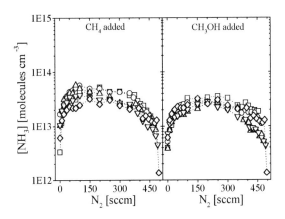

Figure 10.4: Variation of the NH₃ concentration with the nitrogen flow rate under flowing discharge conditions for hydrocarbon precursor admixtures of 40 sccm ($\square$), 25 sccm ($\bigcirc$), 15 sccm ($\triangle$) and 5 sccm ($\triangledown$) and various H_2 / N_2-ratios. For comparison concentrations without hydrocarbon admixture are presented as well ($\diamondsuit$)

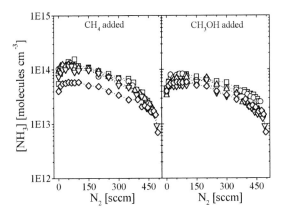

Figure 10.5: Variation of the NH₃ concentration with the nitrogen flow rate under static discharge conditions for hydrocarbon precursor admixtures of 40 sccm ($\square$), 25 sccm ($\bigcirc$), 15 sccm ($\triangle$) and 5 sccm ($\triangledown$) and various H_2 / N_2-ratios. For comparison concentrations without hydrocarbon admixture are presented as well ($\diamondsuit$)

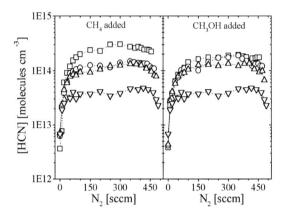

Figure 10.6: Variation of the HCN concentration with the nitrogen flow rate under flowing discharge conditions for hydrocarbon precursor admixtures of 40 sccm ($\square$), 25 sccm ($\bigcirc$), 15 sccm ($\triangle$) and 5 sccm ($\triangledown$) and various H_2 / N_2-ratios

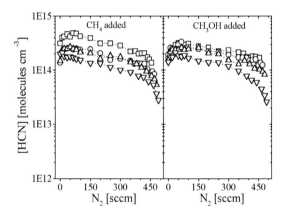

Figure 10.7: Variation of the HCN concentration with the nitrogen flow rate under flowing discharge conditions for hydrocarbon precursor admixtures of 40 sccm ($\square$), 25 sccm ($\bigcirc$), 15 sccm ($\triangle$) and 5 sccm ($\triangledown$) and various H_2 / N_2-ratios

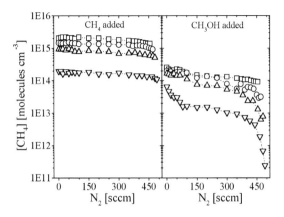

Figure 10.8: Variation of the CH$_4$ concentration with the nitrogen flow rate under flowing discharge conditions for hydrocarbon precursor admixtures of 40 sccm ($\square$), 25 sccm ($\bigcirc$), 15 sccm ($\triangle$) and 5 sccm ($\triangledown$) and various H$_2$ / N$_2$-ratios

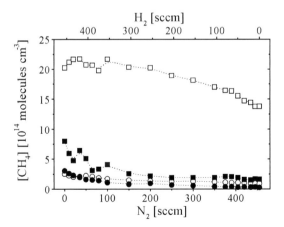

Figure 10.9: Variation of the CH$_4$ concentration with the nitrogen flow rate in hydrocarbon containing discharges under flowing conditions (open symbols) and static conditions (closed symbols). ($\square$,$\blacksquare$ - 40 sccm CH$_4$, $\bigcirc$,$\bullet$ - 40 sccm CH$_3$OH)

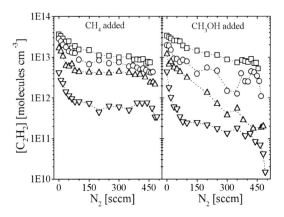

Figure 10.10: Variation of the C_2H_2 concentration with the nitrogen flow rate under flowing discharge conditions for hydrocarbon precursor admixtures of 40 sccm ($\square$), 25 sccm ($\bigcirc$), 15 sccm ($\triangle$) and 5 sccm (∇) and various H_2 / N_2-ratios

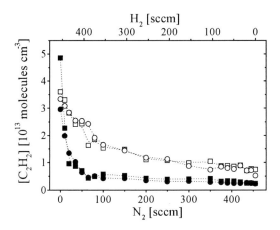

Figure 10.11: Variation of the C_2H_2 concentration with the nitrogen flow rate in hydrocarbon containing discharges under flowing conditions (open symbols) and static conditions (closed symbols). ($\square$,$\blacksquare$ - 40 sccm CH_4, $\bigcirc$,$\bullet$ - 40 sccm CH_3OH)

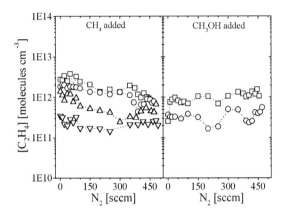

Figure 10.12: Variation of the C_2H_4 concentration with the nitrogen flow rate under flowing discharge conditions for hydrocarbon precursor admixtures of 40 sccm ($\square$), 25 sccm ($\bigcirc$), 15 sccm ($\triangle$) and 5 sccm ($\triangledown$) and various H_2 / N_2-ratios

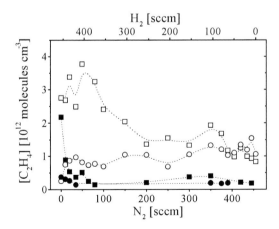

Figure 10.13: Variation of the C_2H_2 concentration with the nitrogen flow rate in hydrocarbon containing discharges under flowing conditions (open symbols) and static conditions (closed symbols). ($\square,\blacksquare$ - 40 sccm CH_4, $\bigcirc,\bullet$ - 40 sccm CH_3OH)

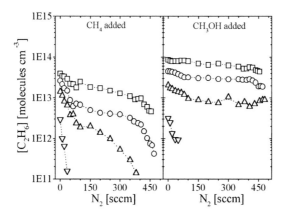

Figure 10.14: Variation of the C_2H_6 concentration with the nitrogen flow rate under flowing discharge conditions for hydrocarbon precursor admixtures of 40 sccm ($\square$), 25 sccm ($\bigcirc$), 15 sccm ($\triangle$) and 5 sccm ($\triangledown$) and various H_2 / N_2-ratios

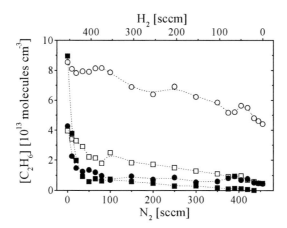

Figure 10.15: Variation of the C_2H_6 concentration with the nitrogen flow rate in hydrocarbon containing discharges under flowing conditions (open symbols) and static conditions (closed symbols). ($\square$,$\blacksquare$ - 40 sccm CH_4, $\bigcirc$,$\bullet$ - 40 sccm CH_3OH)

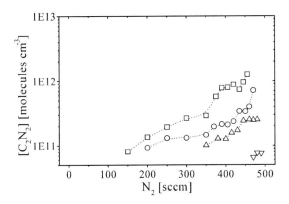

Figure 10.16: Variation of the C_2N_2 concentration with the nitrogen flow rate in methane containing discharges under flowing conditions. ($\square$ - 40 sccm CH_4, $\bigcirc$ - 25 sccm CH_4, $\triangle$ - 15 sccm CH_4, $\bigtriangledown$ - 5 sccm CH_4)

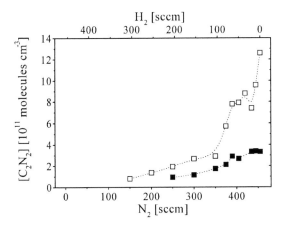

Figure 10.17: Variation of the C_2N_2 concentration with the nitrogen flow rate in discharges containing 7.2 % methane. ($\square$ - flowing conditions, $\blacksquare$ - static conditions)

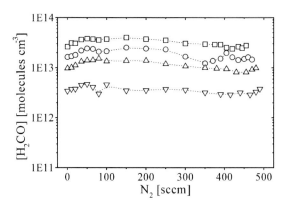

Figure 10.18: Variation of the H₂CO concentration with the nitrogen flow rate in methanol containing discharges under flowing conditions. (□ - 40 sccm CH₃OH, ○ - 25 sccm CH₃OH, △ - 15 sccm CH₃OH, ▽ - 5 sccm CH₃OH)

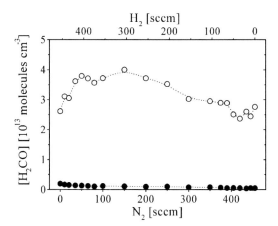

Figure 10.19: Variation of the H₂CO concentration with the nitrogen flow rate in discharges containing 7.2 % methanol. (□ - flowing conditions, ■ - static conditions)

Danksagung

An dieser Stelle möchte ich mich herzlich bei allen bedanken, die mir seit Anfang 1999 bei der Vorbereitung und Durchführung der Messungen sowie bei der Auswertung, Beschreibung und Interpretation der Ergebnisse hilfreich unter die Arme gegriffen haben. Natürlich sind hier viele Namen zu erwähnen; und deswegen hoffe ich, niemanden zu vergessen. Aber trotzdem geht auch mein Dank an alle nicht namentlich erwähnten Mitarbeiterinnen und Mitarbeiter des Instituts für Niedertemperatur-Plasmaphysik e.V., ohne die eine Erstellung der vorliegenden Arbeit sicher nicht möglich gewesen wäre.

Mein besonderer Dank gilt vor allem meinen beiden Betreuern Prof. Dr. J.P.F. Conrads und PD Dr. J. Röpcke für das sehr ergiebige und spannende Thema, für die mir zur Verfügung gestellten experimentellen Möglichkeiten sowie für die sehr gute persönliche Zusammenarbeit im Allgemeinen. Als hilfreich habe ich insbesondere die permanente Möglichkeit von wissenschaftlichen Diskussionen und die Hilfe bei Problemfällen empfunden. In solchen Fällen und bei Fragen in Zusammenhang mit der Absorptionsspektroskopie und der verwendeten Apparatur im Speziellen stand mir auch immer Dr. L. Mechold zur Verfügung, bei dem ich mich auch besonders bedanken möchte.

An der Telefon-Hotline hatte Dr. D. Nelson, Aerodyne Research Inc., bei jedem Anruf Tipps und Hilfestellungen beim Umgang mit der Rapid Scan Software und oft auch eine neue Programm-Version parat. Thanks for your Support on the Hotline, Dave. I hope my feedback was as helpful for improving TDL as your answers were for my measurements. Auch an Dr. P.B. Davies, University of Cambridge, geht ein großes Dankeschön für seine große Unterstützung beim Formulieren der vorliegenden Promotion im Englischen. Thanks a lot Paul for your support, especially for the proof-reading. Für die vielfältige Unterstützung bei der Durchführung der Messungen danke ich unseren Technikern D. Gött and S. Sass. Herzlichen Dank auch an Dr. D. Loffhagen für die Zusammenarbeit und die interessanten wissenschaftlichen Diskussionen in Zusammenhang mit der Modellierung.

Außerdem danke ich meinen Eltern und allen Freunden und Bekannten. Bei ihnen konnte ich immer Unterstützung und ein offenes Ohr bei den alltäglichen Probleme, aber auch Ablenkung vom Laboralltag finden.

Hiermit erkläre ich, daß diese Arbeit bisher von mir weder an der Mathematisch-Naturwissenschaftlichen Fakultät der Ernst-Moritz-Arndt-Universität Greifswald noch einer anderen wissenschaftlichen Einrichtung zum Zwecke der Promotion eingereicht wurde.

Ferner erkläre ich, daß ich diese Arbeit selbständig verfaßt und keine anderen als die darin angegebenen Hilfsmittel benutzt habe.

Tabellarischer Lebenslauf

Persönliches

Name: Frank Hempel
Geburtsdatum: 01.11.1971
Geburtsort: Greifswald
Adresse (privat):   Dubnaring 5 A
                               17491 Greifswald
Telefon (privat):   03834 811 912
E-Mail (privat):   frank-hempel@gmx.de
Adresse (dienstlich):   Institut für Niedertemperatur-Plasmaphysik e.V. (INP)
                               Friedrich-Ludwig-Jahn-Strasse 19
                               17489 Greifswald
Telefon (dienstlich):   03834 554 431
Email (dienstlich):   hempel@inp-greifswald.de

Schulbildung

1978 - 1988   Polytechnische Oberschule „Erwin Fischer" in Greifswald
               Abschluss 10. Klasse mit „Auszeichnung"
1987 - 1990   Lehrgang „English for you" an der Kreisvolkshochschule in Greifswald
1988 - 1990   Erweiterte Oberschule „Friedrich Ludwig Jahn" in Greifswald
               Abitur mit „Sehr gut"
1990 - 1991   Grundwehrdienst bei der Bundesmarine (Stralsund, Peenemünde)

Hochschulausbildung und Wissenschaftlicher Werdegang

1991 - 1997   Ernst-Moritz-Arndt-Universität Greifswald, Studienrichtung Physik
1997   Abschluß des Physik - Studiums mit Diplom, Prädikat „gut"
1997 - 1998   wissenschaftliche Hilfskraft am INP
seit 1999   Doktorand mit dem vorliegenden Thema

Liste der Veröffentlichungen

1. Basner, R., Foest, R., Schmidt, M., Hempel, F. Becker, K., „Ions and Neutrals in the Ar-Tetraethoxysilane (TEOS) RF Discharge", Proc. XXIIIth ICPIG, Toulouse 1997 Contr. IV 196-197

2. Foest, R., Basner, R., Schmidt, M., Hempel, F., Becker, K., „Properties of the RF-Discharge in Ar-Tetraethoxysilane", Gaseous dielectrics VIII, ed. by L.G. Christophorou and D.R. James, Plenum Press, New York, 1998, 161

3. Hempel, F., Foest, R., Hannemann, M., Schmidt, M., „Abschätzung der Plasmaparameter (n_e, T_e) der kapazitiv gekoppelten HF- Entladung mittels Brenn- und Biasspannung", Frühjahrstagung Plasmaphysik der DPG, Bayreuth 1998,Verhandlungen der DPG, 3/1998, 345.

4. Hempel, F., Hardt, P., Röpcke, J., Schmidt, M., „Einsatz der Infrarot-Laser-Absorptionsspektroskopie zur Bestimmung der HCl-Konzentration in der gepulsten Gleichstrom-Entladung in einem $TiCl_4/N_2/H_2/Ar$-Gasgemisch zur TiN-Beschichtung", Frühjahrstagung Plasmaphysik der DPG, Heidelberg 1999,Verhandlungen der DPG, 4/1999, 362.

5. Miethke, F., Gundermann, S., Hempel, F., Röpcke, J., Wagner, H.-E., "Investigations on temporal development and spatial dependence of chemical conversions in the column and the afterglow of low pressure glow discharges in CO_2", 14th ISPC, Prag 1999, M. Hrabovsky, M. Konrad, V. Kopecky Eds., Proceedings Vol. II, pp. 759-764.

6. Hempel, F., Röpcke, J., Mechold, L., "Bestimmung der Ammoniak-Konzentration in einer $H_2/Ar/N_2$-Mikrowellenentladung bei geringfügigen Zumischungen von Methan und Methanol mittels Infrarot-Diodenlaser-Absorptionsspektroskopie", Frühjahrstagung Plasmaphysik der DPG, Bonn 2000,Verhandlungen der DPG, 5/2000, 1014.

7. Hempel, F., Mechold, L., Röpcke, J., „Diagnostics of molecular species concentrations by tunable IR diode laser absorption spectroscopy in H_2-Ar-N_2 microwave discharges containing small admixtures of methane or methanol", XV. ESCAMPIG, Lillafüred , Miskolc 2000, Z. Donko, L. Jenik, J. Szigeti Eds., Europhys. Conf. Abstr. Vol. 24F, pp. 64-65.

8. Hempel, F., Röpcke, J., Miethke, F., Wagner, H.-E., „Infrared absorption spectroscopic studies of the CO_2 conversion on a low pressure glow discharge", ECAMP VII - Frühjahrstagung Plasmaphysik der DPG, Berlin 2001,Verhandlungen der DPG, 5/2001, 177.

9. Hempel, F., Mechold, L., Röpcke, J., „Infrared absorption spectroscopic studies of H_2-Ar-N_2 microwave discharges containing methane or methanol", Frontiers in Low Temperature Plasma Diagnostics IV, Rolduc 2001, Conf. Proc. pp. 92-95.

10. Hempel, F., Mechold, L., Röpcke, J., „Infrared absorption spectroscopic studies of stable and transient molecular species in H_2-Ar-N_2 microwave discharges containing methane or methanol", XXV. ICPIG, Nagoya 2001, Conf. Proc. **4** pp. 233-234.

11. Hempel, F., Mechold, L., Rousseau, A., Davies, P. B., Röpcke, J., „Recent results in TDLAS-diagnostics of hydrocarbon containing plasmas", 3rd International Conference on Tunable Diode Laser Spectroscopy, Zermatt, Switzerland, 2001, Conf. Proc. 44.

12. Mechold, L., Osiac, M., Hempel, F., Röpcke, J., Davies, P. B., „Free radical measurements in plasma diagnostics", 26th Intern. Symp. on Free Radicals, Assisi, 2001, Book of Abstracts, P 77.

13. Hempel, F., Mechold, L., Röpcke, J., "Absorption spectroscopic studies of molecular species in H_2-Ar-N_2-microwave discharges with small admixtures of hydrocarbons", 54nd Annual GEC, State College USA, 2001, Bulletin of the American Phys. Soc. 46/6 (2001), DTP 40.

14. Hempel, F., Röpcke, J., Miethke, F., Wagner H.-E., „Absorption spectroscopic studies of carbon dioxide conversion in a low pressure glow discharge using tunable infrared diode lasers", *Plasma Sources Sci. Technol.* **11** (2002) 266-272.

15. Hempel, F., Röpcke, J., Pipa, A., Davies. P. B., "New laser measurements of the fundamental band of the CN radical in a CH_4/N_2 plasma", Frühjahrstagung Plasmaphysik der DPG, Bochum 2002, Verhandlungen der DPG, 5/2002, 37.

16. Hempel, F., Röpcke, J., Pipa, A., Davies. P. B., "Infrared absorption spectroscopic studies of carbon containing species in CH_4 / N_2 / Ar plasmas", 2002, ESCAMPIG 16 / ICRP 5, Grenoble, N. Sadeghi, H. Sugai Eds., **2**. 133.

17. Hempel, F., Röpcke, J., Pipa, A., and Davies, P.B., "Infrared laser spectroscopy of the CN free radical in a methane-nitrogen-hydrogen plasma", *Molecul. Phys.* (2003), in press.

18. Hempel, F., Davies, P.B., Loffhagen, D., Mechold L., and Röpcke, J., "Diagnostic Studies of H_2-Ar-N_2 Microwave Plasmas Containing Methane or Methanol using Tunable Infrared Diode Laser Absorption Spectroscopy", *Plasma Sources Sci. Technol.* (2003), submitted.